Reflections from the Tenure Track

This book draws on the author's experiences during her 5-year journey towards tenure at a top-tier university in Singapore to offer an internationally relevant, STEM-oriented perspective with reflections, insights, and practical tips for building an academic career in STEM fields.

Often, little is really known about what it means to be a university professor until one is actually in the role, and much learning is done 'on the job'. This book collates and updates blog posts that the author has written throughout her career to explain the multiple roles of an assistant professor at a top research university. It chronicles the day-to-day challenges, pressures, and rewards of academic life. It is organized thematically and covers topics such as what an assistant professor actually does, how to foster an innovative research group, how to achieve target research outcomes and stand out from the crowd, how to create and manage a diverse research group, and how to achieve a sustainable work–life balance. It provides both practical advice and personal reflection, enabling readers to better understand the job requirements and prepare for success within this challenging and competitive environment.

This work provides a vital insider's perspective to early-career academics, including postdoctoral researchers and assistant professors, as well as graduate students aspiring to academic careers. It also contains valuable information for university administrators, educational policymakers, and anyone responsible for career development programmes and academic mentoring initiatives in universities.

Hortense Le Ferrand is an associate professor at Nanyang Technological University (NTU) in Singapore, in the schools of Mechanical and Aerospace Engineering and Materials Science and Engineering. She earned her PhD degree from ETH Zurich in Switzerland before joining NTU as an assistant professor on a tenure track.

Reflections from the Tenure Track

Insights and Advice for Early-Career Academics in STEM

Hortense Le Ferrand

CRC Press
Taylor & Francis Group
Boca Raton London New York

CRC Press is an imprint of the
Taylor & Francis Group, an **informa** business

Designed cover image: Chinnapong

First edition published 2026
by CRC Press
2385 NW Executive Center Drive, Suite 320, Boca Raton FL 33431

and by CRC Press
4 Park Square, Milton Park, Abingdon, Oxon, OX14 4RN

CRC Press is an imprint of Taylor & Francis Group, LLC

Library of Congress Cataloging-in-Publication Data
Names: Le Ferrand, Hortense author
Title: Reflections from the tenure track : insights and advice for
early-career academics in STEM / Hortense Le Ferrand.
Description: First edition. | Boca Raton, FL : CRC Press, 2026. | Includes
bibliographical references and index. | Summary: "This book draws on the author's experiences during her five-year journey towards tenure at a top-tier university in Singapore to offer an internationally relevant, STEM-oriented perspective with reflections, insights, and practical tips for building an academic career in STEM fields. Often, little is really known about what it means to be a university professor until one is actually in the role and much learning is done "on the job". This book chronicles the day-to-day challenges, pressures, and rewards of academic life. It provides both practical advice and personal reflection, enabling readers to better understand the job requirements and prepare for success within this challenging and competitive environment. This work provides a vital insider's perspective to early-career academics, including postdoctoral researchers and assistant professors, as well as graduate students aspiring to academic careers."—Provided by publisher.
Identifiers: LCCN 2025044523 (print) | LCCN 2025044524 (ebook) |
ISBN 9781041063742 hardback | ISBN 9781041227144 paperback | ISBN 9781003635116 ebook
Subjects: LCSH: College teachers—Tenure—Singapore | Science—Study and teaching (Higher)—Singapore | Technology—Study and teaching (Higher)—Singapore | College teachers—Social conditions—Singapore | College teachers—Vocational guidance—Singapore
Classification: LCC LB2335.7 .L44 2026 (print) | LCC LB2335.7 (ebook)
LC record available at https://lccn.loc.gov/2025044523
LC ebook record available at https://lccn.loc.gov/2025044524

ISBN: 9781041063742 (hbk)
ISBN: 9781041227144 (pbk)
ISBN: 9781003635116 (ebk)

DOI: 10.1201/9781003635116

Typeset in Palatino
by codeMantra

Contents

Acknowledgements

This book would not have been written without the encouragement from my parents to transform my blog into another form of reading. It would also not have existed at all without the great support from my students, staff, and colleagues. Thank you to all the students who came to me asking what being a professor is like. Special thanks go to Dr Rohit Pratyush Behera and Prof Koh Khiam Aik (Michael Khor) for their feedback, writing tips, and constructive suggestions.

1

Introduction and Preface

1.1 Reasons for the Book

This chapter introduces the reasons for the book, which are to give some personal insights about what it is to be an assistant professor in a research-intensive university, to break up the mystery about the position, to foster motivation and provide some guidance.

If you have opened this book and are reading these pages, it is likely that you are interested in an academic career, if you are not already in one. During studies, graduate students and postdoctoral researchers may consider the idea of pursuing an academic career as a natural continuation of their academic training. After all, one rarely enters an academic career without pursuing a PhD at least (although there are always exceptions). Especially after turning points like graduation, enthusiasm to continue down the academic path pushes PhD students to enrol as postdoctoral researchers, maybe move to another country or change their research field, in order to build up a CV and experience to obtain a more permanent job as a professor at a university.

In this book, unless specified, professor refers to assistant professor on a tenure-track, which is sometimes viewed as the 'golden' route in academia, likely because during the tenure-track, the assistant professor has to demonstrate that they are worthy of obtaining the tenure. Obtaining tenure means obtaining a permanent position, which ensures job security. The tenure clock starts on day one of the tenure-track and lasts between 4 and 7 years, approximately. To apply for tenure, the assistant professor provides a portfolio summarizing their accomplishments and demonstrating their excellence in research, teaching, and service. That portfolio, or 'tenure dossier', is reviewed by established professors and university leaders.

However, for PhD students and postdoctoral researchers, little is really known about what the actual job of a professor is. For undergraduate students, it is even less known. The mystery might make the position more appealing or instead drive people out of academia for other fulfilling careers. Besides, there are also only a given number of universities and limited professor ship positions available, which makes it quite competitive to reach.

DOI: 10.1201/9781003635116-1

Getting more insights about what the job is can help prepare oneself better for the job interview. Also, when starting as a fresh professor, additional guidance can be welcomed as the learning curve is quite steep.

Furthermore, in the collective imaginary, although it is progressively evolving, a professor tends to be an old white man, which can make the position more attractive to some people who identify with the stereotype and less attractive to others. In order to keep a diverse pool of students and researchers in academia, as well as to help them fulfil their goals down the path of professorship, more knowledge and insights about the job position are needed. Finding mentors able to provide 'behind the curtain' information takes time or may not always be easily available. It is one of the hopes of the book to be a small informal mentor that anyone can reach out to.

Finally, researchers typically apply for assistant professor positions after one or two postdocs of 2–3 years, and therefore, they are well trained as researchers. But there is no training on how to be a professor. As a student or researcher, it is not easy to find clear information on the expectations of what an assistant professor should do and what to do to obtain tenure, as this depends on the field, country, and university, among other variables. Even the obtention of the tenure may also be different depending on the situation in the university. Overall, better knowledge and given examples of what an assistant professor should or could do could alleviate the stress of the position and provide more opportunities to make the right moves. This is thus the aim of this book: to share more about what an assistant professor on tenure-track could do and to give some tips and advice to obtain tenure.

1.2 Scope of the Book

This section presents the details of the scope of the book, including some limitations the reader should be aware of when reading.

This book is specifically about being an assistant professor on tenure-track at a research-intensive top university. An assistant professor on tenure-track is hired on a temporary contract by the university and has to prove themself to become a tenured professor, which is a permanent contract. In most universities, assistant professor is one step down the path to become a full tenured professor at the university, although there may be other steps in between, such as tenured associate professor. There are also other faculty positions that exist, such as lecturers, readers, assistant professor not on tenure-track, or associate professors not on tenure-track, etc. Because this book is based essentially on the personal experience of the author, its content may not apply to other types of professorships than assistant professor on tenure-track, while some insights may still be useful.

This book presents what is to be an assistant professor, but this book is not about how to get into such a position. The process to apply, take the interviews, and negotiate to join such a position is not covered here. There are many other resources that can be found specifically on that topic. However, having a deeper understanding of what the job actually is may help you prepare for such a position.

Furthermore, universities differ from country to country, and even within the same university, schools and departments have different ambitions and requirements. This book is written from the experience of an assistant professor on tenure-track in Singapore, the city-state located under the tropics in Southeast Asia, at a top research university in the Science, Technology, Engineering, Mathematics (STEM) field, more specifically in Mechanical and Aerospace Engineering. Singapore is a very small country that has experienced a booming growth in less than 60 years. In the past two decades, Singapore has been ramping up its research in its academic institutions to drive the economic transformation of the country. As such, the Singapore government has been investing a lot in research at the forefront of many technologies and has equipped its universities and research centres with top-notch facilities and resources. Given its small size and limitations in materials and other natural resources, Singapore also places effort and high value on its people, including schemes to attract international scholars. Singapore's ambitions are to translate the research into real and concrete impacts that can benefit Singapore society in a short timescale, as well as the broader world. Therefore, this focus on research and its impacts, nationally and internationally, is reflected in the book with a strong emphasis on research and less on teaching.

Besides the geographical and cultural context, this book is also written by a woman with no kids who is also a foreigner in that country and who has previous international and interdisciplinary experience. Due to the 'personal experience' nature of the book, there are a number of personal point of views and opinions shared throughout the book. The views, thoughts, and opinions expressed in this book belong solely to the author and do not necessarily reflect the official policy or position of any affiliated institutions, organizations, employers, or other individuals. What is written in the book is obviously not a golden truth, and it is expected from you, the reader, to determine what makes sense to you and to your situation. You may be surprised by some of the content or disagree with some of it. Hopefully, this can make you reflect on your own personal context and find your own approaches to address similar issues or topics.

1.3 The Making of the Book

Here the author shares more about her journey, the motivation for writing the book and how it comes from a collection of blog posts she has written during her entire tenure-track.

Before even applying for an assistant professor position, I attended a number of talks provided at my university, where professors were sharing about their journey and provided very personal insights about how and why they became professors. There was a large diversity of paths, and it was very eye-opening as a PhD student. In my final year, before graduating, I also got the chance to attend a kind of 'speed-sharing' workshop where 5–6 professors (in this case, women) met for about 30 minutes each with a group of 5–10 students or researchers interested in becoming professors. Each session was about sharing their personal path and providing very practical tips. This was very insightful and interesting and made a great impression on me. What was particularly unique is that these professors were sharing personal insights that cannot be found in textbooks. One example I remember was always buying socks of the same colour for the whole family to make it faster in the morning. Another very personal insight was wearing yellow to feel more confident. I believe these insights are hard to come by and are very helpful to better understand what a professor does and what kind of job it is.

After my graduation with a PhD degree from Switzerland, I had the opportunity to do a 4-month research stay at a university in the US in a group that was just starting. This also helped me in witnessing how to start a career. I then moved to Singapore as a visiting postdoctoral researcher under a fellowship from Switzerland. I got a position as an assistant professor in the same university where I did my postdoc. Getting a position in the same university where one has been studying or doing a postdoc is highly unusual. In my case, I think it worked out because I was a 'visiting' postdoc, not an employee of the university, and I changed schools within the engineering domain. While I did my postdoc in the School of Materials Science and Engineering, I started as an assistant professor in the School of Mechanical and Aerospace Engineering due to my interest in manufacturing and processing. After 1 year as an assistant professor, I obtained a fellowship that promoted me to Nanyang assistant professor, which is a more research-intensive position than a 'normal' assistant professor in the university. After 5 years, I obtained tenure and got promoted to associate professor.

I had started a blog during my PhD because I just wanted to write. I continued that blog during my postdoc, and while I cannot exactly remember why I decided to continue during my tenure-track years, I had in mind the idea that I could do something out of it after I get tenure. This book, therefore, finds its content from more than 200 blog posts that were posted on a WordPress website during my entire tenure-track. The link to the blog is: https://hortenseleferrand.wordpress.com/. Having blog posts written ensures that there is genuine and authentic material for sharing in this book, as memory fades and time for writing becomes scarce.

I started organizing the structure of the book by reading all my blog posts and gathering them by topics. I then filled the gaps to cover all topics relevant to the tenure-track which I had not mentioned on my blog. The blog posts were also fairly short and not necessarily relevant to the scope of this book; hence, everything has been rewritten and expanded to meet the format of a book. I also extracted some content from my internal research group website and other personal files.

While the personal sharing part is the primary source of content, other relevant resources and references are added at the end of each section, so that more detailed or complementary information can be found as well. Most of these additional references are papers or books I read when preparing for my assistant professor position and for tenure.

The book is written in a conversational and down-to-earth, practical way, assuming that as a PhD student, a postdoctoral researcher, or an assistant professor on tenure-track, your aim is to get tenure and to do exciting research. Again, I point out that it is intentional for the book to have a personal voice in order to share some insights that may not be found in other resources. It is expected that as a reader, you will make your own decision whether the advice and tips are applicable or not to your context.

1.4 How to Read the Book

Here the author presents the structure of the book and the specific content of each chapter.

The book is divided into chapters and subsections. The titles should give clarity into the specific content of each section, and each chapter can be read separately.

After this Introduction, Chapter 2 depicts the roles of an assistant professor, covering the three main aspects of the job, which are research, teaching, and service, as well as other aspects, such as administrative tasks, communication, awards, and organization. This chapter gives an overview of what an assistant professor does and emphasizes the multiple hats that a professor wears. This is the longest chapter of the book, given its broader scope, whereas the following chapters dive deeper into specific topics.

Chapter 3 is about how to establish and lead a research group. The chapter gives practical suggestions on how to build and organize a physical laboratory, as well as how to create a research group. The chapter is motivated by the research-intensive nature of the position.

Chapter 4 deals with supervision of students and researchers, from how to recruit them to how to manage them. This chapter discusses the practical organization of the research group.

Chapter 5 is about writing grants and getting funds, from the genesis of the grant idea to forming a team, submitting the proposal, and managing the awarded funds. Getting funding is capital to ensure a 'healthy' research group and to conduct research projects leading to scholarly outputs.

Chapter 6 focuses on writing and publishing, covering the different kinds of manuscripts and written media that can be published, and provides some practical writing tips. Scholarly publications are, at the time of writing this book, one of the main performance assessments for obtaining tenure.

Chapter 7 covers teaching at a university. Different types of teaching, as well as advice, on preparing the course materials and the assessment and on conducting the teaching are given.

Chapter 8 discusses work–life balance and includes social media. This chapter presents the challenges in maintaining a balance, what balance means, and some suggestions on how to manage balance in both physical and virtual spheres.

Chapter 9 gives suggestions on how to create a diverse and inclusive research group, and why this matters. The chapter also provides suggestions on how to handle diversity if you are 'part of this diversity', which means being seen or feeling like a minority in your working environment.

Chapter 10 is about the tenure-track dossier or portfolio that assistant professors on tenure-track need to submit to apply for tenure. The chapter gives details about the expectations, the preparation, and how to assemble the dossier.

Chapter 11 is a reflection on the tenure-track process and outcome, including the main challenges and hurdles encountered, the paths to overcome them, and what happens after tenure.

Finally, Chapter 12 gives concluding words with a summary of the book and key learning points. The gaps and limitations of the book are highlighted, as well as final thoughts about the overall tenure-track process.

Each chapter contains personal sharings at the end of each subchapter to give personal, specific examples to better illustrate the content, again with the aim of being very practical. These sharing examples are based on true experiences and are kept concise to be up to the point. In addition, key important points are listed that summarize the content of the section. This is for easier browsing through the content and serves as a take-home message. Finally, each chapter ends with additional references related to its content for further reading.

I hope you will enjoy the book and find it useful.

PERSONAL SHARING

In the past years as an assistant professor, I have been asked many times by undergraduate, master's, PhD students and postdoctoral researchers, what a typical day of a professor is. I usually start by saying that every day is different, then go on with the different things I have been doing recently. I found that undergraduate students are usually surprised to learn that professors do not do only teaching. I was telling one student that writing is the most useful skill as we have to write a lot. Other undergraduate students were inquiring about time management, if it was a 9–6 job, and trying to see if academic jobs could be for them. However, PhD students and postdoctoral researchers mostly inquire about how to get the job and less about what the job is about.

KEY IMPORTANT POINTS

How to read the book:

- Keep in mind the specific context and the motivations of the book.
- Check out the additional references for complementary details.
- Enjoy the reading.

2

The Job of an Assistant Professor

2.1 Overview of the Role of an Assistant Professor

The position of an assistant professor in theory covers research, teaching, and service. In practice, an assistant professor has many more jobs. This subchapter describes the multiple roles of an assistant professor in general terms, highlighting key differences between the job of a researcher, which might be mostly based in a laboratory and focusing on one or a few research topics, and the job of a professor who is now based in an office, but is multitasking with several research projects, teaching, service, and more.

The role of an assistant professor at a top research university is first and foremost to build a research group. Following the US system, the professor will be in charge of their own research and of their own group. This differs from other academic systems where an assistant professor may be joining an existing team under another professor or with several other professors. This scenario is not discussed in this book.

An assistant professor on tenure track faces a large number of new tasks and new roles in comparison to their previous experience as a PhD student, a postdoctoral researcher, or a senior researcher. The primary aim of an assistant professor on tenure track is thus to accomplish what is necessary to obtain tenure, which ensures a long-term position. Typically, it takes 4–7 years (without counting maternity, paternity, or other leave) to prepare for tenure. Obtaining tenure also means getting official recognition of one's skills and expertise, and can make transition between universities easier, as well as open up new leadership roles. Obtaining tenure means it is necessary to know what the goal is and what the targets are to prepare the expected high-quality tenure application dossier. The exact quantifiable key performance indicators for an assistant professor may not be entirely clear as they vary a lot from one place to another, including within one's own university, and can evolve with time. More and more, qualitative aspects also play a key importance. Hence, as a whole, what does an assistant professor do?

There are three main pillars to the job of an assistant professor: research, teaching, and service. Each pillar has different weightages and those weightages vary, even within the same university. In a research-intensive university,

DOI: 10.1201/9781003635116-2

research is typically the most important pillar to focus on. Knowing the importance of each pillar is key as it will guide where the priorities should go. It will also help with defining the expectations and adjust efforts towards excellent performance.

In research-intensive universities, it is typically said that one needs to excel in research and be good in teaching and service. In short, one cannot get tenure with poor teaching but being excellent at teaching without the research excellence will not give you tenure. The service pillar is generally of lower official importance, although it can help a lot with the research, as will be discussed later in the book. However, again, very poor service in the sense of being a poor colleague or breaching some university rules will definitely not be helpful for getting tenure. One needs to keep in mind that getting tenure means *a priori* a long-term job, hence you should show that you will be a wonderful colleague in the university for the next 30 years or so.

Although there are three major pillars that the assistant professor needs to focus on, the actual job encompasses many other tasks such as hiring, budgeting, setting up a practical and physical lab, reviewing and writing papers, grant applications and dissertations, interviewing students for admission, mentoring undergraduates and postgraduate students, delivering talks, marking exams, invigilating exams, etc. One major change from a previous position as a researcher is the large span of new skills that need to be built on in a short time. The other major change is that one may switch from a position mostly located in a laboratory doing experiments, simulations or calculations, to an office desk answering emails, yet multitasking and taking decisions. An assistant professor is expected to become independent and to decide their own research trajectory. That research trajectory may be similar to what had been written in the research statement when applying to the position, or may be different.

Although the tenure track sounds a lot and has some pressure to it, the leadership that is expected from an assistant professor builds up with the years and experience, and plays a great part of the enjoyment of the job. It is a position to make things happen and make people grow. Having the big picture of the profession in mind therefore can help alleviate the stress from the tenure and instead see it with a positive and enthusiastic light. This may be key when things get a little 'bumpy'.

A common regular joy and pride is to graduate students and see them move on to other fulfilling jobs or careers, or even life events like prizes, awards, weddings, births, etc. Because of the teaching and mentoring, the human aspect, meaning the connection to the students, is quite present, and generally enjoyable. PhD students can stay on a project for 3–4 years, undergraduate students may continue as postgrads and their time spent at the university can be decisive to their life or career. Celebrating successes, big or small, embellish the journey and creates fond memories for everybody.

Finally, every assistant professor on tenure track should remember that they got hired by the university in the first place, hence by a committee of established professors who saw potential in them to achieve the tenure and remain as a colleague. Hence, it is also of the interest of the university to support their assistant professors to thrive and become leaders in their field.

PERSONAL SHARING

Despite reading a lot of articles and blogs about assistant professors starting their research group before getting myself into the position, I still found that I had no idea about what the job actually was. Universities are evolving along with the society they are in; I was surprised to find that the job is a constant learning process, as systems and rules evolve.

KEY IMPORTANT POINTS

Overview of the role of an Assistant Professor:

- Research, teaching, and service are the three main pillars.
- Get to learn about the expectations of your university.
- Be prepared to learn to handle multiple diverse tasks.

2.2 Setting up a Research Vision

One key expectation for an assistant professor is to gain independence as a researcher and establish their research group. This subchapter discusses how to manage a research group in general terms and, most importantly, how to find the research direction and give a vision to the research group.

A PhD student generally follows the project topic given by their supervisor. As a postdoctoral researcher, the research project could be also given and set by the mentor and supervisor, or the project may have been already defined by the researcher and approved through a fellowship. Yet, in both cases, the supervisor of the project, the principal investigator (PI) is often the professor head of the lab.

As an assistant professor, there is no one to turn to ask for a project idea. One of the best parts of being a professor and of leading a research group is actually to be able to craft one's own research projects. A key and very important point for an assistant professor is indeed to establish research independence. Often, it is not considered positively for the performance of an assistant professor, and for the tenure track, to be collaborating and publishing papers with past supervisors. From the perspective of the university, they do not want to employ a replica of another professor elsewhere which may make it more difficult for the new professor to achieve high impact. The university wants to have a unique scientist, a leader in their field, or even establishing a new field, an inventor, etc. Hence, it is critical to establish a research vision early in the tenure track, in order to identify what to achieve in the long term.

To find your own 'research signature' and to define yourself as a unique expert in the field, you can start looking at potential groups that you consider could be your competitors. Ideally, you do not want to have competitors close to you, geographically. You can look at the presence of relevant industry which you think could become in the future your collaborators.

A good start to establish the research vision should have been hinted in the research statement submitted when applying to assistant professor positions. The research vision should be refined to have a scope enough for an entire tenure track, which may span 5–7 years, and extend beyond. The research vision should be focused on one specific topic but is also flexible in order to adapt to changes in interest or in funding landscape. One general advice is to be consistent. You should be able to introduce and pitch your work with a few keywords in one or a couple of sentences. You should use the same introduction or pitch when addressing to different people so that you will be remembered for this particular research or area of research. One way to identify your field and where your research lies can be to look at symposia titles at conferences: they often contain catchy yet meaningful keywords. The general research direction should be divided into smaller pieces to determine specific research projects which address fundamental, theoretical or practical scientific issues. Making a list of such projects will help define projects for grant proposals, PhD or master projects, as well as the kinds of papers you intend to write.

Finally, defining a research direction will allow to find collaborators to approach to potentially work with. Having collaborators is generally a must to explore new ideas, extend your knowledge and skills, and adapt to the funding landscape. Indeed, the research direction should be consistent and focused so as you can become an expert known for that specific topic or research field, but should still be wide enough to evolve through the years and to fuel itself from various sources of funding. Also, your personal interests and motivation may shift with the years, and it is therefore natural that research projects evolve along with the career.

PERSONAL SHARING

In the end, I have not been doing what I wrote in my initial research statement that I submitted when applying to my assistant professor position, but I have explored other and broader topics. Looking back at my research statement I emphasized the fabrication of materials for health, in particular implants. This is still one of my interest, but I have been exploring way beyond this specific area, with advanced ceramics for electronics, aerospace, as well as sustainable processes and bio-based composites. The reason I had written about bioengineering was to fit the advertisement. I think it is totally fine I did what I wanted to do. Changing research directions can be strategic but one needs to balance the risks. Starting into a new area may lead to less publications and citations at the beginning, but may lead to higher impacts later. On the opposite, continuing on current research can lead to high publications but may lead to less originality.

KEY IMPORTANT POINTS

Setting a research vision:

- Be consistent.
- Plan ahead for the next 5–7 years.

2.3 Networking

An assistant professor is a job and a title, yet, it does not mean that everything has to be done by the professor themselves alone. This subchapter delves into collaborations and how to fit in the research and funding landscape of one's university and country.

An assistant professor should prove their independence and ability to lead a research group, and be a leader in their field. One way to show independence is the ability to get funding. Studying the funding landscape and identifying potential grant calls, their submission deadlines, the recurrence of the calls, and the format of the proposals is necessary. Most universities have a grant office that can help find these resources. Asking other colleagues or searching the internet can work too.

The initial search for grant calls should ideally be done when applying to a university to ensure that there will be funding available to sustain the research during the tenure track and beyond. In some cases, some flexibility and creativity might be necessary to obtain funds. For example, if your research is very theoretical and fundamental but the grant call is more applied, you may need to emphasize more the application part that the fundamental one in order to be a good fit for that grant call.

It is helpful, if not necessary, to surround yourself with collaborators able to provide the tools, equipment, knowledge, etc. that may be needed to advance your research vision. Collaborations are always a plus, as they enable exploration of multidisciplinary areas. Furthermore, many grant calls have a principal investigator (PI) and a co-investigator (co-I), along with collaborators. In the specific case of grant calls, collaborators generally cannot directly receive funds from the grant while investigators can. Nevertheless, forming a team of investigators and collaborators is generally helpful to craft convincing and impactful grant proposals. Bilateral grant calls also exist between countries, or between universities, which by definition require 2 PIs. Working with other professors as a collaborator or co-PI is useful for grant applications, but it is also fun and motivating to work with other people and to form communities of same interests and goals.

Finding collaborators is an activity in itself and demands some dedication. Conferences and networking events are great places to meet other researchers with similar interests and purpose. Conference presentations give a glimpse into the research capabilities, aims, and style of other researchers. Tapping into past networks and old colleagues and friends can also be another path. Although having some knowledge and trust with a collaborator helps along the way, reaching out 'randomly' to a potential collaborator by interest from their field sometimes can also work, although it may come with some risk. Approaching people from previous networks is easier because making the first contact is simpler. In a new environment, maybe a new country, university, or even a new department, it may be helpful to get a sense of the local culture and how people interact before approaching them for collaborations. The atmosphere between different departments in the same university may be strikingly different, with some departments having their offices always open, while others have office doors closed. Some departments have coffee machines with mingling areas, while others have a networking hour. Finding one's own style for networking and how to tap into available resources is therefore a great start.

Networking with industry is another way to source out funding. Industry time scale is slower than typical university pace and requires time and patience to build trust. Meeting the right and relevant person can be a game changer when reaching out to industry. If there is no existing personal connection, it may be more efficient to tap into university's networking events with industries, the university office for industrial collaborations, or existing industry joint research labs, if any. The challenge with networking with industry lies not only in the time scales, but also in the interests and the culture. The long time scale in industry is generally caused by a hierarchical structure, especially in large groups, hence reaching out to people at the top can expedite things. The other factor is the very different interests of industry compared to university research. Industry has more specific goals in terms of commercialization, returns, and performance. Industry may take a long time to make a decision and be risk-adverse. Industry may prefer incremental research, building on existing processes and materials, while university research prefers more blue-sky research that is highly innovative and disruptive. Industry would also want a concrete solution to be available fast and may prefer investing in shorter projects. The best way to gauge if the potential research done in the laboratory can interest industry is to talk with them at industry events or by other means. Looking for directors of research and emailing them can be a way to have a quick chat. Industry may also reach out to you after your breakthrough research is published and shared in the media. Communicating your research outcomes effectively is therefore necessary. Another option is to approach start-ups instead of large companies. Start-ups have much smaller teams, are less hierarchical, more ready to take risks, and therefore work on a much shorter time scale.

Collaborations give a sense of purpose and reinforce the feeling of belonging to a research field. Collaborations can also lead to more personal connections and collaborators often end up as professional friends, which makes the job even more enjoyable. Joining research societies such as local chapters can help being part of a community, which can later lead to finding collaborators. Joining research societies can be done by volunteering. Most societies are looking for new faces or new people with energy. Joining an alumni group or focussed groups are other types of communities you can tap into. Such groups have the advantage of being more diverse than societies and may also lead to industry collaborations or other types of networking.

Overall, although an assistant professor should demonstrate their independence and own leadership, finding the right people to surround yourself with is necessary to push your research forward and make an impact, while making it more fun and interesting.

PERSONAL SHARING

I find PhD students excellent at finding new collaborators. This happens when they are in need of specific expertise, a chemical, or a piece of equipment. While they reach out through their friends or other student's networks', they evidence the need of joint effort to conduct a certain piece of research, which is what collaborations are for. I have experienced several of such bottom-up ways of building collaborations. In one project, my student wanted to use a new and unique piece of sintering equipment for ceramics and reached out to the other laboratory first to explore the possibility, later asking for my approval. Although I had good connections with this other group, he drove the entire project and collaboration himself. This led to several journal publications and a follow-up grant application. In another case, it was a postdoctoral research fellow who was looking for a way to measure electromagnetic shielding. He found the equipment, talked to the student in charge. I emailed the professor to ask if he was ok. He was. Then the research fellow drove the project, and I met the other professor in person only later.

KEY IMPORTANT POINTS

Networking:

- Find collaborators to strengthen your projects and proposals.
- It is fun and motivating to have a network of potential collaborators.
- Take the time to build your network. Take advantage of existing opportunities.

2.4 Teaching Duties

Teaching is of course one of the key roles of an assistant professor. Yet, professors receive little, if any, training in pedagogy and teaching. This subchapter provides essential tips for effective teaching as well as strategies for effective time management when preparing for teaching.

Teaching is an evident duty of professors. Teaching can be a bliss or can be painful. Generally, one challenge with teaching is that little training and experience has been received during PhD studies and even as a postdoctoral researcher. Also, one tends to base one's own teaching style on one's own experience, yet teaching content and style evolve with time and place. What students were learning 10 years ago is different from what students are learning now. The pedagogical methods are also different, for example with more digital means involved. This evolution and difference in teaching is even further accentuated when changing countries, where the local culture and history greatly affect the teaching methods and teaching content.

Another aspect to keep in mind when facing a new crowd of students, is that if you came all the way to being a professor, it is likely that your relation to academic studies was special to start with. It is crucial to realise that maybe only a handful of students may be interested in pursuing postgraduate studies and continue in research. The motivation of the largest population of students is likely to get a degree and to get a job.

It is therefore good to learn about the students you will be teaching to and to better understand who they are, what it is that they know, and what it is that they want to learn and for what purpose. Several ways to achieve this include talking to colleagues, asking to attend their classes, reading papers in the field of pedagogy, or volunteering for events involving pre-university students, such as judging or outreach events. Learning about the different levels of educations and the different paths leading to university is also helpful.

The university has probably a program or available resources to help train new professors to teach effectively. It is highly recommended to see how to leverage these resources to reinforce your teaching. It is always good to be reminded of the basics and also to be provided with useful practical tips adapted to the university. Universities are places for learning, and this includes professors. Having a growing learning curve is always viewed positively, hence it is ok to use the first classes you teach at the beginning as training and to improve from them. Practicing teaching always makes it better.

Preparing course material can be daunting too. When preparing the material, taking examples from other existing courses helps in gauging the right amount of material to be taught and how. If possible, try to take over an existing course where you can tap into existing teaching material. If you are teaching an entire new course, you may want to look into existing course material online to combine them together and to gauge the depth of teaching that is appropriate for your students. Having an existing reference book for teaching the content can be extremely helpful for you and for your students. When delivering the course, you should be enthusiastic while teaching the material else it will be difficult to interest and to engage students. Find your own style to achieve this. Also, the course material will need to be revised regularly or updated, due to changes in course schedule, co-lecturer, or other improvements. One trick is to avoid placing dates, names and pictures of lecturers on the teaching slides, keeping another slide deck for these inputs. This reduces the number of slides that need to be updated from one semester to the other. Courses may be taught with other lecturers and demand some coordination and maybe homogeneity. It is good to be clear on the schedule a few weeks before the start of the semester.

The courses to teach may be assigned to you or you may be able to choose them or even propose them. It is possible to be given a course on a topic that is new to you, which may demand some preparation for you to learn about it first. If possible, you can ask about the course topic and also which semesters you want to teach. It can be strategic to teach only one semester so that you can attend conferences in the other semester, for example. It is also good practice to look at the academic calendar a year in advance to mark down exam periods, recess week, or other important dates. If missing a class because of travel, you will need to make some arrangements so that the students have the class. To accommodate these events, you may change the teaching schedule, for example, so that another co-lecturer teaches that day, or you may be able to provide a video recording instead, etc. Be aware that there may be part-time teaching, which you might be asked to do, and that likely will occur in the evenings. It is recommended to discuss with the person in charge of attributing the teaching load to faculty if you have requests or want to make some modifications to your assignments. Because teaching is a must but not the most critical part of the tenure track, finding ways to make your life easier is a good idea.

Another daunting part of teaching is the teaching assessment. Find out early how your teaching will be assessed. It may be based on feedback survey, in which case knowing which questions will be asked to the students will help you design and conduct the course to meet these requirements. If it is based on peer review it is also very good to attend classes by other colleagues, especially those with good teaching reputation, in order to get a sense of what is expected and to learn effective teaching practices. Define what a good teaching is for you as well and do some reading about it. This will be useful later for writing the teaching statement in your tenure track dossier. Mentoring and supervision are also part of the teaching, and will be discussed more thoroughly in the relevant section.

PERSONAL SHARING

I have been teaching manufacturing for undergraduate students using power point slides in a lecture theatre. The first times I taught, even with the addition of videos, I found that it was very difficult to interest the students, and myself as well. I found a way to enjoy my teaching by doing small experiments in class. For example, I used gallium, a metal that can melts at 30 °C, to showcase the low viscosity of liquid metals by melting it in my hands. Not only it becomes very fun for me and for the students, with a little bit of adrenaline when it takes a bit too long to melt, but the connection to real objects supports the content and facilitates the memorization of the key concepts I demonstrate. I also bring a 3D pen that extrudes thermoplastic polymer through a handheld nozzle, like a 3D printer, to compare the viscosity between plastics and metals.

KEY IMPORTANT POINTS

Teaching duties:

- Learn about your student's knowledge and expectations.
- Be aware of the evolution of teaching content and expectations.
- Optimize your lecture preparation to accommodate for this constant evolution.
- Ask about teaching schedule and evaluation in advance.

2.5 Optimizing Service

Service is often pointed out as a minor activity for assistant professors. However, service can refer to various types of activities which can be very rewarding to career and research, when chosen carefully. This subchapter gives some inputs on how to choose and conduct effective service, with concrete examples.

Service can range from sitting in a school's committee, being an examiner of PhD or master theses, to peer-reviewing papers, grants, or organising symposia and workshops. Typically, as an assistant professor, service requested from you should not be too much. However, service can actually be a great tool to learn, or even help your research. For example, examining student presentations and PhD defences is a great way to get to know the work of your colleagues, as well as to better understand the standard expectations for the students in your university. 'Small' service activities such as being a judge for hackathons or helping with undergraduate student's clubs can help a lot to understand the student's population, how the university works, and getting trained on asking questions and giving feedback. Because all these will take time, it is important to choose activities or events that are fun and interesting to you. You may also consider volunteering for activities you find specifically interesting and useful to you.

Service can help to give your research some visibility and provide with pathways to form connections with various research communities. For example, joining a research society or a local chapter of a bigger society can be a form of service, with roles such as committee member. These may not demand too much work and can be opportunities to organize symposia at small or major conferences. Starting as a co-chair of a symposium is a great way to learn the ropes, then taking on more responsibilities in the next years. Not only it is an opportunity to meet people, but you will also be prompted to send emails to colleagues to invite them and your name will be visible on the symposium or conference website, hence raising your visibility. Organising a symposium also will increase your chances to be invited to organize another one or to be invited in return by your invited speaker to give a presentation at another event. Service done can also often be shared on social media, which can also increase the visibility and the chance to meet new collaborators.

Participating in conferences can also be a form of service, and will definitively help disseminating your research. Funding the travel and registration to attend conferences can sometimes be challenging. Asking for discounts to the conference organisers surprisingly works a lot of time, especially as early career professor. Being invited, of course, is always a good way to increase the visibility while adding a line to your CV, but it can also be rewarding to choose a conference to present in order to meet a new community, for example, in a specific region or about a slightly different topic than the one you have been working on so far. Making the most of a conference means listening to the talks and networking. It can also be very helpful to take notes, in particular, to note the names and affiliations of remarkably inspiring speakers. Typing down these notes and names and storing them in a folder with date and place can be surprisingly helpful later, especially 5 years later, when looking for potential reviewers for your papers or student's theses, for inviting speakers in your symposia, or even for the referee list of your tenure package.

Taking on paid consultation for companies or other institutes can also be a rewarding form of service as it demonstrates expertise and impact. Since it may take time to build industry connections, consultation service may arrive after a few years only. Part time teaching or any other special program teaching may also be a type of consultation and service but may require a lot of time and energy. Hence, that type of consultation would be recommended only for topics you are very passionate about. Peer-reviewing grants is also a type of service that is good to demonstrate expertise and impact, especially industry-related grants or international competitive grants. It also helps in identifying what makes a good grant proposal.

One way to optimize the service is therefore to pick up service that can be useful in some ways to build a stronger CV and tenure package. And if not, that is at least, fun.

PERSONAL SHARING

During my tenure track, I did a lot of symposia and workshop organization. Although this took time, it was instrumental for me to keep in touch with my network of collaborators and to create new connections as I relocated in another country. These initiatives were helpful in later getting conference invitations and finding collaborators for international seed grants. For example, we have access to a number of small funding to organize a 1 or 2 days workshop between 2 international universities. Organizing one enabled me to share my network with my overseas collaborator and vice-versa. Later, I had submitted grant applications with some of the invited speakers of our workshops. Similarly, I had been invited to their own workshops too.

KEY IMPORTANT POINTS

Optimizing service:

- Service contributions should be chosen carefully to support your goals.
- Volunteer or pick service contributions that can be added to your CV or that are meaningful to you.

2.6 Administrative Tasks and Procedures

The amount of administrative tasks to be conducted by an assistant professor is often underestimated before joining. This subchapter describes what these administrative tasks can be and how to optimize them.

As a student or researcher, we may operate within one 'small' and 'closed' environment with a colleague and a supervisor. This means that your direct environment, with which you are interacting, is first and foremost your research group. As an assistant professor, you get to talk and interact with a whole range of other people, including laboratory technicians and managers, people from human resources, finance, communication, all with various degrees of seniority and expertise. All these interactions are related to various administrative tasks.

The amount of administrative tasks is difficult to realize until you are in the job, trying to make something happen. Take the exams for example, a university may be dealing with thousands of students and has strict rules to avoid chaos from happening. Hence, it is very likely that your exams questions and answers will be checked and verified several times, by several people, to ensure there is no mistake. Also, exams are confidential, which makes the sharing of the papers involving either tons of passwords and internal folders, or in person handling of documents to manually sign. Hiring is also likely a new topic for you, especially hiring researchers like postdocs. The university will also have to follow the country's rules perhaps with a set number of guidelines. Sometimes, it may be challenging to hire a very good person because of VISA issue. Dealing with all this can be confusing at times, also given that rules may change and vary by factors outside of your control. In addition to examination papers and hiring researchers, administrative tasks can include grant reports writing, submitting papers to the open access university repository, doing grant variations to shift the budget from consumables to manpower, signing consultancy papers, approving purchases, writing reports of PhD defence, updating one's CV for annual performance assessment, etc.

The key to make administrative tasks seamless is to learn to ask very clear black and white questions in order to receive very clear instructions on how to make things work. It may also be good to realise that sometimes, a negative answer to something may just mean that the question was not clear enough, or that there was a language issue. Any rejection or being asked to resubmit that or that form should not be taken personally. Each administrative segment may have their own terminology and language, and understanding it can be a game changer to expedite things. The structure of administrative tasks is also a lot more rigid that what is usually encountered when doing research in science and engineering, as there is a need to follow protocols and policies. Despite being time and energy consuming, and sometimes painful, administrative tasks are safeguards as they avoid any involuntary breaching of rules, mistakes, or anything else that could have more serious consequences in the end.

Because it can be daunting (sometimes, submitting a simple claim for a business trip can be a mine field), it can be helpful to schedule a time, when you are in good mood and not in a hurry, to do them. Often, it is easier done than was anticipated. When you are the one in position of giving approvals, for example, to purchase chemicals or consumables for your lab, they can be grouped together instead of doing them one at a time. Depending on the hierarchy of people checking every document, sometimes, it can be safe to approve quickly knowing that other people will also have to check anyways... It is key to be aware of the administrative timeline to avoid delays. Having the chance to meet and know the administrative teams in person can make the process a lot nicer and friendlier. It can also help to find who is the most efficient person or who has that particular knowledge to help you.

PERSONAL SHARING

I find it very helpful to be able to place a face to the admin person I am communicating with. As many, I dislike these tasks, but realizing there is a person on the other end helps smooth things out, especially when my requests cannot be processed because of a mistake, missing document, missing information, or other. It also helps to understand what and why we need to do that particular task. A little kindness is also always well received from the other end and often make things happen faster.

KEY IMPORTANT POINTS

Admin tasks and procedures:

- Administrative tasks include communicating to a wide range of diverse people, not only researchers and professors.
- Learn to ask clearly about instructions. Check the deadlines.
- Realize that administrative paperwork is useful and needed to run a university properly.

2.7 Outreach & Communication

Despite not being often highlighted as important, outreach and communication activities can be critical to get recognition and for standing out from the crowd. As an assistant professor, this can make a big difference for promotion and tenure. This subchapter discusses about these activities and points out their benefits while suggesting ways to carry them out effectively.

Outreach means reaching out to the broader community, to the public, or to future students. Outreach may not seem to be so important at first, when working in academia, where publications in scientific journals and conferences are the most important. After all, professors are not media communicators, journalists, or advertisers. But successful outreach can demonstrate impact and can be quite rewarding for a professor, as the positive effect also shines onto the professor's team, the university and even the funding agency. Outreach will increase visibility of your research and allow to reach to a larger number of people than only the academic circles. In a simplified picture: Universities want to have superstars that will bring shine to their institutions. International recognitions and international rankings are often assessed by how many people know about the university and what positive image they have from it. Hence, having faculty shining to the general public is very rewarding for the institution.

In STEM fields, outreach can be about sharing a result from the research, such as a discovery or an innovation, but can also be sharing about your work through a workshop, an exhibition, etc. Although the most impactful would be to have the broader media reaching out to you because of the impact of your work, this may not happen by itself so often. Therefore, one needs to be proactive when doing outreach and may require some little practice.

Outreach for beginners can start from answering one's emails and responding positively when companies, potential collaborators, or future students want to visit your university or laboratory. Giving a laboratory tour is an easy skill to acquire and it is usually always a pleasure to show what is the great work you are doing in your lab. This can lead later to having students applying to join your research group, or word-of-mouth dissemination about your expertise. Social media can also be an easy way to do outreach although consistent effort and strategic choices may be needed to make the most impact without too much time. One issue with social media is how quick things change as new platforms open, while others lose popularity. Beyond lab visits and social media, there are many other ways to do outreach, and at various scales. Examples can include conducting workshops into schools to motivate and interest younger students to pursue studies at the university. Appearing in the news can also be achieved by informing the communications office of the university about a major upcoming paper. If there is funding available, for example, as a grant is finishing and there are remaining funds, it is possible to find magazines or journals where pages for sharing news can be purchased. Outreach can also include writing a blog, participating in a public exhibition or seminar. Outreach activities can take various forms and there are activities for all tastes. One can look for opportunities or create the opportunities to suit your own agenda.

Doing outreach means sharing about your work and/or yourself under a bright light. It often generates positive feelings and although it may seem futile, it can be quite useful and rewarding. The usefulness comes with the new skillset, which is mostly to communicate to a wider audience, in writing or orally. Writing and presenting skills are always needed for publications, grant applications, among others. It can also be introspective and help reflecting on your current research, for example, if you have a blog to share about your thoughts. It can also help understanding better what key challenges certain communities face and therefore help in proposing impactful solutions to address these challenges.

Some recommendations about outreach are to remain professional and to do it with a specific purpose in mind. Ideally, you want to control what is shared about you and not being in the news for something you are not proud of. Using artificial intelligence may not be too recommended, at least needs sufficient editing to avoid coming across as someone who just uses AI to share. The amount of outreach and the type of outreach may also be carefully chosen, spread in time and including various types of outreach. For the same reason that sometimes being too visible can be detrimental and make other people feel uncomfortable, it is important to find a just and adequate balance in the amount and content of outreach that you are conducting. One advice to boost outreach visibility is to include pictures of people, including yourself, in the shared content. Pictures are catchy and always lead to the most positive feedback and re-sharing. Tagging the relevant people, institutions, and adding links also help increase the visibility.

PERSONAL SHARING

One example of positive and useful outreach activities I participated in was to be featured in podcasts where scientists talk about their research. I initially found an advertisement about it on social media and volunteered for it, then I got invited into several others. Not only it is always pleasant to talk about the research we do in the group, but I always find it inspiring to listen to other's questions and ideas too. I enjoy listening to podcasts myself and find them inspiring. The podcasts being open to public on platforms like Youtube, I found out some people knew about me through these media, especially students, which is extremely rewarding and also tells about the interests and motivations of the students. I also found being interviewed in podcasts a very good activity to practice sharing about technical research in simpler words and to emphasize the impacts that I want my research to have. I find this activity contributing to the shaping of my overall research vision.

KEY IMPORTANT POINTS

Outreach & communication:

- Reaching to the broader community can set you as a leader in your field and be a key advantage to your tenure package.
- Outreach can be nurtured proactively by responding to invitations or sharing about your work with your university media department and social networks.
- Choose outreach method and media with a specific purpose.
- Remain professional at all times.

2.8 Tips on Organization & Planning

One challenge when juggling multiple roles is to stay organised and manage everything effectively. This subchapter gives concrete practical tips on organisation and planning.

Starting as an assistant professor can be overwhelming, not only because of the variety of new tasks to conduct, but also because of the sudden change of 'timescale', meaning now you should be thinking and planning for the next 5 years at least. Just for your application to get tenure, you will be putting up a dossier summarizing all your activities and achievements of the past years since you started as assistant professor. Organization, therefore, is critical to not lose track of any information and keep comprehensive records of your activities.

Organization and careful planning should also extend beyond the scope of the job as an assistant professor in order to optimize your productive time. This means that you should also consider carefully your own personal goals and work-life balance when organizing your work. Given the marathon-nature of the tenure track, identifying and securing time for work and time for rest and other personal activities is very important. Academia is flexible on the working hours, which is a unique privilege. Identifying which working pattern fits your needs can be very helpful. For example, if you are working best in the morning, scheduling your writing in the morning can be a good strategy. The work-life balance chapter will delve deeper into this topic.

There are multiple ways to organize your calendar. Current electronic tools enable seamless planning and organization. Whether you like a flexible or a more rigid timetable, having some planning and a semester and yearly overview of what to expect is very useful. Planning and booking time for holidays a year in advance may seem extreme, but invitations for conferences also come 6 months to one year in advance and not having carefully booked time off for vacation may lead to you saying yes to conferences and not being able to find 1 or 2 consecutive weeks free of commitment to take a break. It is good to be aware of this. Because commitments pile up, either for research, teaching or service, blocking your calendar for critical activities in advance will avoid delays and stress. Critical activities refer to grant submissions or lecture preparation and exam marking. Indeed, these activities have clear deadlines and cannot be rescheduled. Prioritising the research outputs and manuscript submissions is also a good idea for your tenure track.

In addition to your own tasks for your research group and teaching, there might be a number of events and meetings happening at the school, college, university, or even at the broader community level. Quickly identifying which ones to prioritize is also a good way to avoid spending hours in not-so-useful meetings. For example, it may be possible to inquire if there will be a minute of the meeting sent afterwards. With new AI tools, meeting organisers can generate such minutes automatically. How to decide which meeting to attend can vary. Sometimes, one needs to attend first to understand how that particular type of meeting is getting conducted. Sometimes, knowing who is attending or presenting at the meeting can be a determining factor.

Organization and planning also means to have a clear overview of the tasks to carry out in the future. To-do lists are of course one easy way to remind oneself what to do and to prioritize certain tasks. Using digital tools can automate the process with simple tools such as flagging emails to work on later. There are many methods on how to stay organized and to optimize time and effort, such as time blocking, thematic scheduling, etc. Studying them can be a time-saver.

For the tenure dossier, you will have to provide proofs of invitations, awards, any recognition, students' notes, etc. that you would have received during the tenure track. It is therefore critical to keep a track record of everything and to store them in a way to find them easily. The track record should be kept up to date regularly. For example, 5 years later you may be requested to provide a screenshot of a conference website to show that you were an invited speaker, but the conference website may have been deleted. An example of good practice could be to save and export every document of potential importance in PDF and to save them in clear folders and making regular back-ups of the folders. This is further detailed in the chapter on the tenure track dossier.

Keeping a clean and organized computer desktop is also helpful for all other activities. There is no point in saving documents if they cannot be found later. It is therefore worth to spend time establishing a folder system suitable to your organization style. An example of folder organization can be:

- Conferences
- Students reports
- General administration
- Grants
- Students and researchers
- Manuscripts
- Meetings
- Outreach
- Teaching
- Trips

Finally, with the digitalization, most of documents and important communications are carried out by emails. Hence, it may be desirable to organize your emails effectively as well. The number of emails received can be relatively large. Universities likely will have a spam filter, but it is good to develop a habit for recognizing emails of low or higher importance based on their sender and subject. Some emailing boxes also have tools to help with your organization, such as flagging emails or attributing a colour to them. It is a good suggestion to look and try out some of these tools.

Technology also now calls for multiple diverse online platforms for communication, including Whatsapp, Wechat and so on. It is helpful to be flexible in order to communicate with various colleagues and other stakeholders, but it also may be a good idea to also have a central organization. Back-ups should also be made regularly, using physical back up or on various digital clouds. Using these various online sharing methods also calls for caution about the safety and confidentiality of the documents shared and stored. It is good to check the university policy or recommendations on where to store which kind of data, to prevent any hacking, leakage, or any other issues. Following good digital communication ethics and housekeeping with regular changing of passwords and back-ups and cleaning non necessary data is also a good idea.

PERSONAL SHARING

Because storing all these data takes a lot of space, purchasing data storage on digital platforms like iCloud was useful. I also use the research data storage from the university to store large research data (which I only discovered recently when writing these words). I find it helpful to create folders starting with the date and information about the content to be able to find it many years later. I also export PDFs of important emails because I find it too difficult to find old emails years later.

KEY IMPORTANT POINTS

Tips on organization & planning:

- Get organized from the start to prepare your tenure dossier.
- Book your calendar in advance to have at least a 6 month overview.
- Practice comprehensive record-keeping in well-organized folders.
- Find an organization system for storing your digital data.

2.9 Checking Progress & Aiming for Awards

For an assistant professor on tenure-track, the first career goal is to successfully obtain tenure before the end of the tenure clock. To ensure that the progress is moving according to plan, there is a need to regularly check one's own advancement. This also enables to visualize the full picture of the work for crafting the statements in your tenure dossier and applying to awards.

As an assistant professor on tenure-track, you will need to put up a tenure dossier to submit to be considered for tenure. The tenure dossier comprises a full and detailed CV, the research, teaching and service statement, a statement about future work, and a proposed list of referees. The dedicated chapter 10 on the tenure track dossier later in this book gives more details about the tenure dossier. To be able to put together such tenure dossier, one needs to have an overview of what has been done and what one intends to achieve. We discussed in the earlier sections the importance of record-keeping and planning the research vision at the short and long term. Therefore, planning ahead the 5 years of the tenure track with clear short term and long-term goals will definitely be very helpful.

One suggestion for planning ahead the tenure track is to prepare a PowerPoint (PPT) presentation about the current and future research projects and interests. Such a PPT can be handy in having a visual representation of what has been done, and what you plan to do. It can also be showcased for future collaborations. The PPT format can also be easily edited. A template for such a PPT could be:

- Slide 1: What is the domain of application of the research? For example, any grand challenges target?
- Slide 2: One sentence describing the overall research question.

- Slide 3: One sentence describing what is the overall aim of the research.
- Slide 4: Lists 3–4 research focus areas.
- Slide 5: The research approach: collaborations, affiliations, key methods.
- Slide 6: The research capabilities such as equipment, methods that make the lab unique.
- Slide 7: The overview of the current (or intended) research projects under the key research focus areas.
- Slide 8 and further: the details of the research projects, indicating the title, the funding sources, the papers published or to be published, the invited presentations, the technical disclosures or patents, the outreach events conducted, the collaborations and the names of the staff and students who participated to the project. One visual image can be added to best illustrate the project. What has been achieved can be written in black while what is intended to be achieved can be written in red and highlighted in yellow.

Such a PPT as presented here can help visualize the targets (such as publication, patent, invited presentation) as well as how the targets will be achieved (what funding and who exactly will do the work). The targets should be somewhat related to what is expected from you by your university, and can be found out by simply asking your department Chair. It is very important to understand what is valued by your school and university, and to make it match as much as possible with your own core values and intended goals. With regards to who will do the work to achieve these targets, it will help to be able to identify what kind of positions are required: can an undergraduate, postgraduate, postdoctoral fellow or other do the work? What should be their background or skills? With regards to the funding opportunities, it will be helpful to study what funding sources could be a good fit to conduct that particular type of work.

Although such PPT, or similar document, breaks down the long-term research vision into smaller shorter goals, one suggestion is then to narrow down the research even further by listing potential research papers or studies that each available or future student or researcher can work on. Such overview of the research will also help identifying potential awards to apply to or what you want to achieve to apply for an award. Finally, having a concrete visual representation of your work and key performance indicators (the KPIs) also helps boost confidence at times when things may not go as planned, while also being flexible on what and how to achieve these goals. Such PPT should be revisited regularly, such as every semester or every year, and can be shared informally to your team members to convey a sense of working together towards a common goal, while the achieved KPIs can be shared with colleagues or potential collaborators and used for outreach and visibility.

PERSONAL SHARING

Setting clear goals of what I wanted to achieve was useful for my research projects as well as for applying to awards. I usually revisit my goals every 6 months or so during the break between the semesters. Identifying which awards I would like to get and what I would like to have achieved in order to feel that I deserve the award was a huge motivator and enabler. Applying for awards was a way to set oneself as a leader in one area which is a positive card to have when applying for tenure.

KEY IMPORTANT POINTS

Checking progress & aiming for awards:

- Have a clear overview of what you want to achieve and what you have achieved.
- Find out the expectations and set clear targets.

2.10 Conclusions

This chapter gave an overview of the multiple aspects of the job of an assistant professor, based on the three main pillars that are research, teaching, and service. Research should be conducted so as to demonstrate independence and impact. Having a clear long-term vision as well as short-term goals is needed to create a storyline for your work. Research includes applying to grants and being known in your community. Teaching should be approached strategically to be delivered successfully, and service activities should also be chosen carefully to support the research or teaching goals. Organization, clear targets, and knowledge of the expectations should be obtained as early as possible.

Additional References

L.R. Larson, L.N. Duffy, M. Fernandez, J. Struts, J. Gray, G.M. Powell, Getting started on the tenure track: challenges and strategies for success, Schole 34 (2019), 36–51.

Hay, *How to be an Academic Superhero: Establishing and Sustaining a Successful Career in the Social Sciences, Arts and Humanities.* Edward Elgar Publishing, Cheltenham, 2017. https://doi.org/10.4337/9781786438126

Burroughs Welcome Fund and Howard Hughes Medical Institute, Making the Right moves, A practical guide to scientific management for postdocs and new faculty (2006), Second Edition. https://www.hhmi.org/labmanagement, Accessed 27 March 2018.

R.E. Rohde, What could possibly go wrong on the maddening 6-year interview?, Tips for success on your path to tenure. 17 Sept 2015 https://www.elsevier.com/connect/tips-for-success-on-your-path-to-tenure, Accessed 02 May 2018.

L.M. Jarvis, A year in the life of a new professor, inside the exhilarating, frustrating, messy transition to being the boss. 2016 https:/cen.acs.org/sections/year-in-the-life-of-a-new-professor.html, Accessed 06 March 2018

R. Nagpal, The awesomest 7-year postdoc or: how I learned to stop worrying and love the tenure-track faculty life (2013). https://www.scientificamerican.com/blog/guest-blog/the-awesomest-7-year-postdoc-or-how-i-learned-to-stop-worrying-and-love-the-tenure-track-faculty-life/, Accessed 13 March 2018

P.W. Droz, L.S. Jacobs, FAST Professor: strategies for surviving the tenure track, ALRA 2 (2018), 6. https://digitalcommons.humboldt.edu/alra/vol2/iss1/6/

D.A. Verrier, Perceptions of life on the tenure track, NEA Higher Educ J Thought Action 9 (1994), 95.

D.R. Gosling, N.M. Chae, J.R. Goshorn, The tenure-track life: experiences of new faculty in tenure-track positions, William & Mary Educ Rev 7 (2002), 4. https://scholarworks.wm.edu/wmer/vol7/iss1/4/

3

Establishing and Leading a Research Group

3.1 The Physical Lab

The first practical aspect of starting an independent career is to start a research group. This research group, in STEM, usually comprises a physical laboratory. Examples of laboratory settings are provided to suggest how this laboratory should or could look like. Setting up the practical laboratory in the university and who is involved are discussed. The results, advantages and happy surprises of having a lab running are described too.

3.1.1 Organizing the Physical Space

Whether your work is based on theory and computational simulation or on experiments, you will need to build a physical lab where your students and researchers can conduct their research. Depending on the situation, you may inherit some lab space from a previous faculty member who retired or left recently, or someone who does not need the space anymore, or be given a brand new lab space. Whatever you are given, you will need to make it your own, which may demand a bit of re-organization. The amount of renovation that you may be able to do will depend on your own resources, needs, and the support from your university. You can talk to your laboratory manager or the person in charge of the laboratory organization in your department to understand better how this will work. It may also be common for PIs to share laboratory space. This can be convenient for having access to specific equipment. Having one lab just for your own research can also present advantages, such as when taking pictures or videos or giving lab tours to visitors as you will know exactly what can be shared and what is going on in the lab. Identify what the situation is for you to make the most of it.

A simple and intuitive way to represent one's own laboratory space before actually moving in is to review the laboratories that you have previously worked in. You can take some notes about what you liked and disliked in these labs, and what would be useful and what would not be useful for your own lab. Then, you can compare what your ideal lab looks like to what you could possibly have in your new university, given the available resources.

DOI: 10.1201/9781003635116-3

Some laboratories have researchers sitting just next to their lab benches, which is smooth for conducting experiments but may not be desirable if toxic chemicals are involved. Some other laboratories have a clear separation between the laboratory and the office spaces, where the laboratory is divided into different sections depending on the equipment and type of work that is being conducted. This could facilitate maintenance by assigning a group of people to be in charge of very specific parts of the laboratory.

Take note of small facilities that can make a difference. For example, having a white board to make schematics or write messages, or placing storage for consumables next to the equipment needing these consumables. Talk to your laboratory manager to review the university's rules in laboratories, such as distance between lab benches, number of people in a room, etc. They will be able to set up the basics, and you will be the one to ensure it will be functional. This means you should be able to anticipate what your lab will look like in the next 2–3 years. Likely, the number of people in your laboratory will increase and they will need space and many new equipment for their research. Again, the global research vision is necessary to be able to do these projections.

PERSONAL SHARING

I had long discussions with the managers about how to best position the lab benches to include the largest number of people authorized given the space, while ensuring easy motion and travel of people inside the lab. Along with the lab benches came the position of the fume hood which depends on where the exhaust system is and how to squeeze in some storage cabinets. I found that involving my current students and researchers in the organization (only a couple of them) facilitated the process of setting up a new lab, as they could tell clearly what they need. As a foreigner, having local students talk to the local technicians and managers also made everything smoother.

KEY IMPORTANT POINTS

Organizing the physical space:

- Work with technicians and managers to understand the rules and policies.
- Be flexible and creative.
- Remind yourself of the previous laboratories your worked in.

3.1.2 Running the Space

Each laboratory and university have their own procedures, which may change and evolve with time. You can look for equipment from other professors or shared equipment from the department's laboratories. You may want to do a bit of research to understand the costs associated with using shared equipment and how to use and collaborate with other professors. If there is no available catalogue of existing equipment in the university that you can have access to, it can be time- and energy-saving to send students who need to use this equipment to look for it. Students have a lot of resources and university knowledge through their pool of friends and students' clubs. Once the student finds the appropriate laboratory or professor, you can liaise directly with them.

Safety is key in any working place. You may need to take training just for sitting in an office and adopting an appropriate posture. Likely, every experiment conducted in a laboratory will need to have its own risk assessment, standard operating procedures, proper labelling, etc. Make sure everything is conducted in order. Learn about the policies for purchasing, importing, and storing chemicals. For safety reasons, certain laboratories or departments may not be allowed to get certain chemicals, and you will need to find adjacent strategies to be able to use such chemicals. For example, by requesting a bench space in another laboratory, or finding a collaborator that would allow your student to use their lab space, etc. Be aware that managing the laboratory, which includes waste disposal, lab coat cleaning, repairs, may fall into your own grant budget. If you are planning to purchase some equipment, finding space for that equipment and the tender procedure may be new to you as well. Be prepared that it may take 6 months to get the approval for purchasing that piece of equipment. While being aware that there is a lot of background work to make anything happen, you can explore strategies to find and have access to what you need to conduct your research.

PERSONAL SHARING

I assign each student with responsibilities in the laboratory: training and maintenance of equipment, chemical repository and labelling, purchases, waste disposal, audit preparation, and risk assessment. Often, I have two students on each task so that they can help each other or catch up if one is missing. I also assigned one researcher with the processing of the purchases for the entire group. I think it is important to assign responsibilities else it often turns out some people do everything while others do not contribute.

KEY IMPORTANT POINTS

Running the space:

- Follow the safety procedures for each laboratory.
- Learn about the procedures to manage the laboratory and conduct the experiments.

3.1.3 Challenges and Surprises

Building your own laboratory is exciting but may have other unforeseen challenges, such as communicating with technicians and managers, finding common language, and understanding the university's rules. Having students and researchers helping is definitively an added value. Also, be aware that people tend to be protective of their space and equipment, given the effort and budget needed to set it up. Sometimes, open and empty space may not be free to be used. Finding alternative approaches to making things work might be needed, and flexibility is the key. Finally, besides the laboratory space, you may also need to find places for your researchers and students to sit, which may also be in high demand and require some organization. Again, being able to envision how much space will be needed and when, can help you book and reserve the necessary space. Informing managers in advance will be helpful for everybody and allow buffer for organization. Keep in mind that you need to balance between what you need and want, what is possible with the rules, and what is available.

The laboratory space is never static, and there will be unexpected events occurring. Hopefully not major hurdles, although incidents may happen, such as broken glassware, dirty equipment, or mixing of chemicals. Some unusual things to consider may include the effects of the external weather and how it can affect the laboratory temperature and humidity, as well as air composition. When your lab space is set up and running with students and researchers, it becomes very rewarding to see your lab being alive, on its own. A lot of little surprises can sprout here and there, such as small organization boards, funny names attributed to equipment, small intriguing experiments running, etc. (Figure 3.1).

FIGURE 3.1
From left to right: unusually high humidity of 94% in the laboratory; funny labelling of glassware; unusual experimental trial by researchers.

PERSONAL SHARING

Each year 7–10 undergraduate students join the lab to do experiments and each time there are issues such as contamination of chemicals or infections of our cell cultures. While there is always a risk that this happens, simple protocols can be put in place to diminish these events, such as proper training, putting chemicals in smaller containers, etc.

KEY IMPORTANT POINTS

Challenges and surprises:

- Find space for your researchers and students in the lab to do experiments as well as in the office to sit.
- Protect your space.
- The weather or other external factors may have effects on your laboratory.
- A laboratory running on its own can lead to surprises.

3.2 The Lab Spirit and Atmosphere

A research group is not only a physical laboratory, it also includes researchers and students. This subchapter highlights the importance of creating a cohesive research group as well as practical management tips for the research group activities, in and outside the laboratory, such as during group meetings.

3.2.1 The Importance of Your Lab's Atmosphere

The laboratory is not only the lab space and pieces of equipment, it also includes the students and researchers. More specific details and tips on mentoring and supervising students and researchers are provided in the next chapter. Here, the research group atmosphere and spirit are highlighted. Indeed, having a good research group atmosphere may be helpful for attracting students, producing good researchers and research, and making your work known. It is also just more pleasant for everybody.

Your students and researchers are ambassadors of your research group and will be an additional interface between you and colleagues, other students, the public, etc. Universities are increasingly international, and the members of your group may have different standards, different goals and perspectives on how a research group should be. You may therefore want to set clear standards and expectations to make sure the group is according to your desires and wishes.

If you remember your time as a student and researcher, you probably have heard 'stories' about abusive professors, or groups where students were unhappy and where the research was challenging. You may also have experienced practices which you may not agree with, or would have wished things to be different. You know that students and researchers talk to each other, but also to your potential collaborators. Although nobody is perfect and it is never possible to please everybody, it may still be possible to try your best to at least follow the principles you agree with. It is very rewarding to hear from students that they enjoy working with you and in your group.

PERSONAL SHARING

When I was looking for a PhD position, I visited one lab where each student told me how there was competition between each member because each student had to conduct 3 different projects independently. Only when they have sufficient results can they talk to their Prof who then write the paper for them. Every student I talked to advised me not to join that group. I definitively did not apply for the position in that group. I do not want anyone to think of this from my own group.

KEY IMPORTANT POINTS

The importance of your lab's atmosphere:

- Your students and researchers are ambassadors of your research group and of your research.
- Review your past experience as a student and research to visualize how you want your research group to be.

3.2.2 Bonding Activities

Forging a strong and bonded research team may require some creativity. For example, it may not be possible to find funding for conducting bonding activities but voluntary participation, and simple group activities like walks, lunch or coffee can be enough. Asking your team to come up with ideas is already a group activity. They will likely come up with surprising and interesting ideas such as lab decorations, tours, etc. Bonding activities are very helpful to create ties, to get to know more about the personalities of each of your team members, and to evaluate the level and degree of communication needed. Bonding activities can also be the opportunity to take group pictures to show at the end of your presentations in conferences and can be added to your group website. The frequency of the activities you want and can participate in is, of course, your choice. Once a year at least can be a good start.

Because your team might be diverse, it is important to make sure everybody is on the same page. Your research team is like a little enterprise. Having an on-boarding process explaining how the laboratory works, introducing the organization, where the physical lab is and other rules of conduct is recommended. Assigning a buddy to help settling in can also increase the ties within the team members.

In addition to the personal and group meetings further described in the next section, emphasizing internal collaborations and assigning specific lab management roles to each group member can help them take responsibilities and make the lab their own. Some equitable repartition of the tasks is needed to ensure everybody participates. You can give students tips for making booking calendars for equipment, and how to get organized, especially when you are just welcoming the first group members. It is possible that new students or even researchers are also not familiar with operating in a new lab where the rules and guidelines are not yet present. It can take a few months before everything rolls on its own. Letting the team organize itself after a while will also reinforce the cohesion.

PERSONAL SHARING

I could not find funding for group activities like group lunch or workshops. Instead, I noticed I could get consultation money for doing some part-time teaching and student supervision, which I do not mind then investing in running one group activity per year. In the last few years, we have been going for indoors ice-skating, diner, making plant composition, bowling, escape game, etc.

KEY IMPORTANT POINTS

Bonding activities:

- Be creative to find small funds to pay for it.
- Let your team decide and organize.
- Bonding activities can also be research oriented.

3.2.3 Creating a Cohesive Team

In addition to bonding social activities, you can reinforce the feeling of cohesion in your research team by sharing achievements regularly and give yearly reports on achievements, expectations, and where you are heading to. Sharing the overall vision of the lab can help your team understand and visualize the goals and aims of their work and appreciate how it is part of a larger project. This kind of sharing can bring extra motivation and enthusiasm beyond the completion of a thesis or publishing one research paper.

Cohesion and team bonding are also helpful in ensuring a safe and inclusive atmosphere which can in turn lead to more collaborations between your team members, leading to new research ideas and directions, increased productivity and wellbeing for everyone. You can take inspiration from the townhalls or other group meetings you attend as a faculty member, where general vision for the coming year and past achievements are celebrated. This kind of small sharing sessions can include a presentation with a few slides on:

- Overview from the past year: who joined or left the team; what new equipment or facilities have been used; what conferences or events some team members attended; what publications and their impacts; what technologies have been developed; what funding has been received; which industries were interested to collaborate; other special achievements from the team.

- What is happening in the coming year: who will join or leave; what new equipment or facilities will be used; what research direction or new project.
- Areas of improvement: point out what you think did not work properly or what you would like to change and provide suggestions and indications of how it should be changed.
- Organization: any need for reassigning tasks within the team members; find dates for team bonding; other things to discuss.

Such sharing session can be an opportunity for you to ask feedback or suggestions, discuss issues or problems to be solved. Informing the team about the key performance indicators (KPIs) you need to achieve from your grants can also be a way to prepare them for their future positions in academia or elsewhere. It is also an opportunity to say thank you and share your enthusiasm in your research and how their work is meaningful.

PERSONAL SHARING

Students are really good at getting discount or finding alternative solutions to get chemicals or equipment, especially if it has been shared with them that there are restrictions on the budget. When joining a cohesive and bonded research group, sharing about the group plans make everybody feeling part of it and helping each other. I think a certain level of transparency is needed for everybody to understand where they are going.

KEY IMPORTANT POINTS

Creating a cohesive team:

- You can decide how you want your research team to be and tell them.
- Positive lab spirit and atmosphere can motivate and create a productive environment for research.

3.2.4 Managing Conflicts

Despite efforts in creating a pleasant research group atmosphere, conflicts and competition may arise with the group members for various reasons. Communication, prevention, and transparency are key to resolve conflicts.

Inter-personal conflicts may happen and be may not so easy to spot, and can lead to sabotage or resignations. One approach to avoiding this is to assign each person a specific independent project first. Then, if these persons are supposed to collaborate with each other, setting the collaboration with them being present and look for solutions. If conflicts take too high proportions, it would be better to reach out for help to the human resources or other appropriate office at the university.

Other conflicts may come when accessing critical pieces of equipment when the research group becomes large with more than 10 members. In this case, it may be needed to set priorities according to projects deadlines, or to decide of a maximum number of hours per person. Authorship can also be a source of tension, which is critical to discuss and clarify very early on.

PERSONAL SHARING

One interesting source of conflict I had were two postdoctoral research fellows working on different topics, but using similar methods. One research fellow felt that by using one of the method, the other one would steal the novelty of his own paper. Because I had told them during their on-boarding process to tell me if there is anything they are unhappy with (and he told me this is the reason he raised his concern to me), he informed me early and we went into another direction with the other research fellow. This was a good turn of event. One tendency that I see in fairly large groups, is that people start to do the same things, such as using the same software, or doing the same kinds of analysis and tests.

KEY IMPORTANT POINTS

Managing conflicts:

- Discuss early on.
- You can also decide and set rules.

3.3 Research Aims & Practices

Research is the most important focus of an assistant professor in a top research university. It is therefore worth to stress again the importance of having the research vision and to discuss ways to make short and long-term impacts. This subchapter includes discussion about patenting and the applications of the research, sharing the research with the community and with stakeholders, as well as getting in the news for more broader impacts.

We discussed earlier the importance of having a short-term and long-term visions of your research. The short-term vision can be narrowed down to scoping topics for PhD projects, master theses, bachelor final-year projects, and other short projects that students can conduct, as well as larger projects for postdoctoral researchers. All these projects are like pieces of the puzzle that will be assembled together to bring your long-term vision to life. The assembly of these pieces of the puzzle may vary as new project ideas come in. Even with some flexibility, having clear goals helps in communicating them to your team which will then contribute effectively in realizing them. Depending on your supervision style and the attitude of your team members, you may organize your lab in a structured way or not. For projects requiring expertise and knowledge, for example, for operating a piece of equipment or conducting a simulation, it may be helpful to adopt a slightly hierarchical organization where postdoctoral researchers help PhD students who supervise master and bachelor students. However, it is also important to leave space for the students and researchers to also explore their own ideas which can then enrich your own research vision. When starting a research laboratory, having a postdoctoral researcher can be a great help for the installation and organization of the practical laboratory as well as setting standards for students.

Further narrowing down the exact scope of the projects for each team member will ensure that there is no overlap between the projects, while there might be internal collaborations or combinations of research projects together for making up one journal paper. This narrow view can make you scope down the publications that you want to write, therefore helping you guide the researchers and students in this direction. While planning for publications, high-impact publications in high-impact-factor journals are of course the target, as set by current standards in academia. However, heading for smaller journals with smaller scope can first help you publish when you are just starting, but also help build confidence in the students when they start and make them learn how to write a paper. Students' projects, especially undergraduate or master projects have little impact on the student's graduations as they are usually not expected to publish. Hence, these kinds of projects can be good to try new ideas or explore new collaborations with other professors or industries.

Along with journal publications, scoping down the technologies that you plan to develop is good to do early. Indeed, technological disclosures have to be filed before publishing the journal paper and before sharing the results in a conference. Technological disclosures (TDs) may take up to 6 months to be filed. Working closely with your university patent office will help you understand the process. At the start, just submitting any kind of TD can be a way to get into the process and to know the people handling them and what is the procedure, even if no TD is filed in the end. Although your university innovation centre or patent office may help reach out to industries to licence your patents, you can also contact industries at forums or through emails to understand the current needs. Preparing a pitch and a short slide deck can help the communication.

Other short-term goals can include participation to workshops, conferences and exhibitions, entering competitions, and other outreach goals you may seek for. Impacts can be demonstrated by getting into the news and getting the research outside academia to the general public or the specific stakeholders, companies or other agencies.

PERSONAL SHARING

I like to group undergraduate projects together with two or three students tackling different aspects of the same project. If the results are good, the data collected can be used for writing a publication. The students also generally like to work in groups as they can help each other. For example, I had one project on 3D printing of gelatine-coffee-fungus. One of the students looked at the printing parameter using one specific ink composition, while the other one was tuning the composition of the ink and measuring the viscosity. When I started as an assistant professor with only two PhD students, having the undergraduate students doing some research was instrumental for having publications in the first year.

KEY IMPORTANT POINTS

Research aims & practices:

- Narrow down your broad research vision into smaller pieces: students projects, specific publications or technological disclosures.

3.4 Getting Funds

Getting funds is necessary to hire students, researchers, purchase the basic necessities for running the laboratory, from consumables to access of equipment, as well as to attend conferences, etc. This section gives some ideas of where and how to source funding, some strategies, and how to budget effectively.

3.4.1 Grant Proposals

Funding can come from a variety of sources and like for investments, it is a good idea to vary the sources of funding. It can come from internal or external grants, national or international grants, awards from foundations, fellowships, industry, etc. Applying for grants is time-consuming and generally involves a lot of administrative paperwork. It is a good idea to review what grants you can apply to and when the calls open and end. Many grant calls require the use of digital platforms for submission which may possibly come with technical hurdles. It is never a good idea to plan last minute submissions. Also, many institutions have their internal deadlines where approvals should be sought before the final application.

Although it is appealing to have a lot of money, it can also be wise to plan the grant applications strategically to avoid having all grants starting and ending at the same time. In an ideal situation, you always have one grant on-going and one under review. The grant review process can take from a couple of months to a year. Large grant calls may have multiple application steps with interviews and presentations to defend your project. As a rule of thumb, there might be about 1 year between the time you press the submission button to the time the money is seen in your university account. The success rate for each grant call may also vary significantly depending on the call, the year, and luck.

Grant proposal writing is a new task that assistant professors should master. Proposals are written to answer very specific calls. It is, therefore, necessary and important to tailor the proposal to that particular grant call. Study the description of the calls and the requirements. After writing the first draft, place yourself in the shoes of the reviewer and find the answers to the following questions:

- What is the significance of the research? What are the impacts?
- What is the topic addressed by the research? Is it a major topic and is the proposition novel?
- How it is different from other existing works and what are the desired outputs?

Depending on the call, you may want to check which technology readiness level (TRL) will be reached at the end of the research. Industry grants or more applied research, will aim for higher TRLs. It might be a good idea to give specific benchmarks for your desired outputs, with quantifiable numbers to which you can compare your research. It can be the value of a material property, a price, a time, etc. The proposals are also expected to be innovative, but might not be too innovative in the sense that it is good to show preliminary data that prove the research is realistic. The preliminary data can come from your own previous work or ongoing research. Previously published works may have more value to the proposal as they have been peer-reviewed.

Some proposals can have a single PI, or invite Co-PIs and collaborators. The degree of involvement of the team may depend on the funding that each team member may receive and personal interests. As a rule of thumb, PIs who do not receive direct funding may wish to participate by being invited to present and host a researcher, for example. When starting as an assistant professor, and even later, it may feel strange to ask people you do not know if they are interested in joining in. Actually, it is good to dare as it can lead to pleasant surprises. Researchers and professors, in general, are interested in doing research and are pleased to talk about their work, especially with other early-career researchers. Indeed, some calls, especially international calls, may require PIs from different countries and at various career levels. Whether the proposal is successful or not, the proposal writing is interesting and leads to growing your professional network as each member brings their own connections. Because grant proposal writing is time- and effort-consuming, careful selection of the time and which grant call to apply to is helpful. Planning for recurrent grant calls can help prepare in advance too, while some grant calls may be one time only.

Where to start when writing a grant proposal? A good idea is to review the recent and old literature around the topic, while narrowing down to the specific aspects that you wish to tackle. Reviewing the literature to ensure being up to date in the field can be tedious if it is a new area. Using online search or artificial intelligence tools, or attending seminars on the topics can help pinning down the main challenges in one area.

If you have access to previous grant proposals, it can be helpful to read them to see the style and the content, which may vary for each grant call. Some proposals can be found online, as seen in the additional reference lists of this chapter. You may ask some colleagues for tips and old proposals, although this is trickier. Some funding agencies may also run workshops to introduce in more detail their grant calls, in which case it is a good idea to attend the briefing. These kinds of meetings are also good avenues to meet potential collaborators interested in applying to the same grant call with similar topics. Agencies may share specific details that may not be officially written in the grant call, for example, if they are aiming at a specific TRL or have a preferred topic.

If your previous proposal has been rejected, you will likely receive feedback with the comments from the reviewers, which may help you improve the proposal if you are invited to resubmit. Sometimes, giving a call to the agency can help getting informal feedback, which is always useful.

Industry funding can be another avenue for getting money to support your research. Industry funding can take more time to source as it often includes strong economic components and requires trust to be installed with the company. The requirements and KPIs may be different, with less possibility to publish in scientific journals but a greater emphasis on patents and commercial output. If you have no contacts, reaching out to industry can be liaised through the university's established connections, forums, joint research laboratories, etc.

PERSONAL SHARING

I find the best time to write up my next grant proposal is when my existing project is about halfway, and the first results have been obtained. At that time, new ideas usually emerge and it is helpful to write the proposal as some research is on-going on similar topic. If the proposal is unsuccessful, what has been written can be used in a research paper or in a review. If the proposal is successful, then I already know where to start and who could help or be hired on the project. As I am writing this book, I have a new research project starting where one part of the research is actually almost done and finished as we started writing the project about a year ago and the researcher started working on that topic with other students.

KEY IMPORTANT POINTS

Grant proposals:

- Plan your grant applications wisely.
- Diversify your sources of funding.
- Plan early to build your proposal and to submit it.

3.4.2 Budgeting

Budgeting properly is important for running your lab. You likely should provide a budget when submitting your grant proposal and the exact details of your budget can be scoped down and corrected later. You can assume that the first budget you will prepare will not be 100% good as it is difficult to estimate the costs for consumable use and facilities for your research when you have no previous experience. The first grant you have can help to find out how much was actually spent in your first year, facilitating the next budgets. Luckily, budgets are not carved in stone and there is always the possibility to move funds around, although it requires some paperwork.

Adding collaborators possessing one piece of equipment can avoid budgeting for that equipment, while adding maintenance fees and salary for student assistants can be helpful to overcome equipment breakdowns and cover PhD students after their finish their scholarships. What important things to budget vary and asking colleagues can help you foresee what to budget in and strategies that can be good to know.

Avoiding overspending can spare money for manpower. For example, universities have a lot of materials and consumables available, and faculty retiring or leaving the school may leave some basic and useful equipment like lab scales, stirrers, etc. These may be advertised or not. Asking laboratory technicians and other students can help find these out when they are needed. It is good to have your existing students and researchers purchase or source what they actually need to avoid buying equipment that nobody uses. Sharing the budget with your team, at least for the consumables and equipment can be helpful in that matter. You would be surprised at how good they may be at sourcing cheaper deals. At the same time, not spending enough may also not be good, which is why appropriate budgeting is necessary.

PERSONAL SHARING

On my first grant, I bought a big piece of equipment that nobody had at the university and that I thought would be of great use for the project. That specific piece of equipment (high temperature thermomechanical analyser) took effort to buy, install, and in the end only one student used it one time, at which the equipment got damaged. Further maintenance was needed to make it functional again, but it is still barely used by my team. From this lesson, now I avoid budgeting for any large equipment. I budget only a few small equipment and make variations if another piece is needed for the project and is requested by the students or researchers. For using larger or specific equipment, I try to collaborate instead, or outsource the testing.

KEY IMPORTANT POINTS

Budgeting:

- Your first budget will help you shape the subsequent budgets.
- Manpower generally takes up most of the grant money.
- Optimize your budget.
- Budget variations are generally possible: check the rules for the grant you have been applying to.

3.5 Conclusions

Establishing and running a research group means arranging its physical and 'mental' spaces. Laboratory arrangement and organization are key for successful research and can be optimized by involving the students and researchers from the group. Creating a good 'mental' space for the group should also be crafted intentionally by activities to create cohesion, such as team bonding or sharing your vision of the lab. Each lab member should feel responsible for the overall group success and contribute actively to its proper functioning. The research should be organized in clear individual projects. Finally, a 'healthy' laboratory has resources coming from grants.

Additional References

M.R. Monroe, Advice for scientists starting a lab. 29 April 2014 https://blog.addgene.org/advice-for-scientists-starting-a-lab, Accessed 26 March 2018.

M.T. Brown, Preparing and managing your first lab budget: finance 101 for new investigators. 22 Oct 1999 https://www.science.org/content/article/preparing-and-managing-your-first-lab-budget-finance-101-new-investigators, Accessed 01 Aug 2019.

B. Goldstein, P. Avasthi, A guide to setting up and managing a lab at a research-intensive institution, BMC Proc 15(Suppl 2) (2021), 8. https://doi.org/10.1186/s12919-021-00214-7

University, meet industry: 4 ways faculty can make and keep strong industry partnerships. 13 March 2018, https://venturewell.org/blog/industry-partnerships/, Accessed 2 May 2023.

T.C. Erren, P. Cullen, M. Erren, P.E. Bourne, Ten simple rules for doing your best research, according to Hamming, PLoS Comput Biol 3 (2007), e213.

Available grant proposals:

Successful NSF (National Science Foundation) grant proposals, Carleton College. 28 June 2006, https://serc.carleton.edu/NAGTWorkshops/earlycareer/research/NSFgrants.html, Accessed 27 Aug 2019.

My (successful and unsuccessful) ERC starting grant proposals. 20 Sept 2018 https://figshare.com/articles/journal_contribution/My_successful_ERC_Starting_Grant_Proposal/7110767, Accessed 18 Oct 2018.

4

Supervising Students and Researchers

4.1 Why Supervising and Mentoring are Important

Supervising students and researchers can be a challenge for a junior faculty and comprises one important component of the teaching part of the job, while enabling the actual research to be done. This subchapter stresses the importance of supervision and mentoring for these aspects as well as for the general career vision, and joy, of an assistant professor.

As an assistant professor, you will have to recruit students and researchers to join your research group and to carry out research under your supervision. Your group may comprise undergraduate students, master students, PhD students, postdoctoral researchers, and maybe other types of researchers such as research assistants or project officers which are researchers with a bachelor's degree or a master's degree only. Supervising and mentoring them will be crucial to ensure that they align with your research goals, while also developing their own ideas and their own CV.

It is very rewarding to see one's students and researchers grow and flourish. It is also expected from a professor in a university to enable such flourishing. Mentoring and supervision indeed can fall under the teaching category in your tenure track dossier, and show your skills in encouraging people upwards, while the results from their research will contribute to achieving your own global research vision.

Efficient mentoring and supervising are particularly important for PhD students. They already chose to join your team and are placing of a lot of trust in you and usually come with a lot of hope and enthusiasm. PhD students are deeply motivated by conducting research and often join in with initial ideas of continuing in academia. Hence, successfully mentoring PhD students can really help with generating new talents, as well as pushing your own research forward. Graduating a PhD student may also be a written requirement for your promotion and tenure.

Undergraduate students may be joining your research group as well for small projects, and be less involved in your overall research vision for your group. Yet, mentoring and guiding undergraduate students that express interest in

DOI: 10.1201/9781003635116-4

research and in pursuing higher education may also be worth the time as they might become your next PhD students. Mentoring undergraduate students in general can also be fun if you have the time. They often come with a broad range of projects and ideas, which may be short-term but exciting for them. Being able to find some time to provide advice for them can give you a good reputation among the students and in the school, which is always a plus to have.

Depending on your interests, mentoring schemes may also be arranged for pre-university students. This could be a great opportunity to do something different and to impact future generations, merging your teaching and research activities with service. Undergraduate and pre-university students may also be very enthusiastic to take on unconventional research projects, or just to try out some ideas without too much worry about the outcome. This can be an opportunity for you to try out something new and risky.

Mentoring and supervising postdocs is also something new for an assistant professor. But mentoring postdocs comes with the advantage that you might have been in their position not too long ago and can more easily relate with their experience. Postdocs have very different aims than students. Many hope to transition to a permanent or tenure track position in another university and their expectations are not only publishing great papers, but also to build experience in teaching, organising symposia, and any other opportunities that can advance their carriers. Most postdocs are very capable and proactive. Helping a postdoc to continue as professor is also extremely rewarding and can lead to long-term professional friendships and collaborations.

To better understand the university's style and expectations about mentoring and supervision, it is a good idea to participate or attend to the thesis defense or qualifying examinations of students, to look at webpages where projects are advertised, or look up some of your colleague's research group webpage. In addition to reflecting on your own experience as a mentee and student, this can help you create your own supervising style.

PERSONAL SHARING

Supervising an undergraduate student on a project remotely (she was in another country), I could explore one research I wanted to do for a long time, which consisted in creating morphing materials using common and safe ingredients like clay, starch, and plant fibers. Actually, she did the entire research at home! The project lasted only a couple of months and we could write a short paper published in a peer-reviewed journal. A year later, I met a professor in another country who told me how this paper was useful in one of her projects. I would never have conducted this project and written this paper if I had not had this undergraduate student interested in doing some unusual research and putting trust in my online guidance.

KEY IMPORTANT POINTS

Mentoring and supervising:

- Can help boost your research.
- Be very rewarding.
- Impact future generations.

4.2 Recruiting Students and Researchers

How to recruit students and researchers is the topic of this subchapter. The need to understand their point of view for choosing your job position is important. Being able to adapt the project and the tasks to the student or researcher is equally key in ensuring their success. Practical tips for CV screening and effective job advertisement and recruitment are shared.

Recruiting students and researchers will be needed to build your research group. Recruiting the right students and researchers can be crucial to ensure the research goes in the right direction and that you obtain the desired outcomes. The first step for recruitment is therefore to clearly identify what project you would have, what kind of positions would be needed for conducting that project, and what you expect from your new recrew. Likely this will have been set in the research proposal already, but clarifying it in smaller bullet points would help communicate clearly to candidates what is expected from them.

Besides what has been proposed in a grant, there might be a lot of other various opportunities to embark students or researchers into one project. For example, there might be exchange undergraduate or postgraduate students, compulsory final year projects, master dissertations, etc. If you identify clearly various projects with defined scopes and activities, it will become easier and faster to advertise these projects broadly. All projects can be redefined when someone takes up the project as the research may have moved on. You may also be lucky to have applicants with their own funding, such as postdoctoral researchers with their own fellowship or PhD students with their own scholarship. How they would fit into your research agenda is to be thought of before they join your research group to ensure they will not dilute your research focus, unless you are also open to explore other areas.

4.2.1 General Guidelines for Recruitment

You will need to find how to advertise for jobs through your university. The job advertisement should provide the important information without being too specific. It is a good idea to keep it a little open so as to adapt and tailor the project to the applicant. Sometimes, a good applicant shows up who could fit better in another topic, then there will be an easier opportunity to adjust accordingly. Some candidates may apply because they are interested in the specific project whereas some candidates may apply because they are interested in joining a young PI or they have heared from you by other students or researchers. It may be interesting for PhD students to join an assistant prof with a smaller group as there might be more interactions and direct support. Students also may look at your awards or past experience and want to join for these reasons. Postdoctoral researchers may like more experienced professors to get access to a network but may also like a younger PI to get more discussions and direct support too. Since there are many factors that may prompt someone to apply, it is good to keep the application open and to inquire further during the interview if the CV shows the competencies that you are looking for.

Before starting advertising and selecting candidates, it is good to inquire about hiring policies. There may be internal process or regulations related to salary, nationality, education, ... that need to be followed. These kinds of regulations may change regularly without being broadcasted. Candidate screening and selection can be time-consuming, having a preferred strategy to sort them out can be helpful. Example of preferred strategy may be to first ask colleagues, or instead post the job advertisement on social media platforms like LinkedIn, etc. Adopting strategies when reviewing CVs is also helpful. For example, you can get yourself a list of the renown universities in your region or look at specific competencies. Asking a colleague to look at a CV can be worth to have a second opinion, especially when you are new to the area or you want to know the reputation of a certain university from another country.

After screening the CVs, interviews and requesting for reference letters need to be conducted. Interviews take time, so reducing the number to 1 or 2 candidates only may be more efficient. For the reference letters, if there is possibility to directly call or email the referee, you may receive more specific and spontaneous recommendation or information. Reference letters tend to be all the same and to be non-specific. A good practice for interview is to ask the candidate to conduct a short presentation that includes an introduction of themselves, their current research, and their future plans. Inquiring further on their motivation for applying is critical to know how dedicated and enthusiastic the candidate is. As recommended for the job advertisement, not mentioning the very specific details of the project at first allows to tailor a little bit the discussion after they have shared more about their expertise, especially if this is an outstanding candidate but with slightly different interests. You can also ask them more about how they heard from your research, if they know the place, and other specific questions to ensure they would fit well within your research group. For example, if you are looking for someone to collaborate with another student or researcher, do they have any experience working in teams? Also, if the candidate is from another country, do they know your country or anyone around? When people have to relocate internationally, the time to adapt to a new context may vary for various reasons. Giving opportunity for the candidate to ask questions is also appropriate to let them share any other information you might need to know.

Choosing the right time to advertise for a job position can be tricky. If the grant has started and you start looking for candidates, it may delay the research. If you look for candidates before the grant starts, but the grant start is delayed, the candidate may look for another position in between. It is particularly tricky when there is a good candidate you would like to have, but you know your grant is starting a bit later. Overall, nevertheless, there are more candidates looking for jobs than there are positions available.

PERSONAL SHARING

I was informed by a colleague that an excellent PhD candidate was defending soon and looking for postdoctoral position. The candidate was open to relocating internationally and wanted to pursue in a research topic that exactly fitted in one of my upcoming grants. I was advised to arrange a discussion with that future researcher, and they would be great for the position. Yet, the grant start was not yet confirmed at the time and there were some delays in starting the recruitment, leading to telling the candidate that we could not meet their timeline. In the end, they found another position and were not interested anymore later, and we found another good candidate too, from a local university.

KEY IMPORTANT POINTS

General guidelines for recruitment:

- Be aware of recruitment rules at your institution.
- Be strategic as it can be time-consuming.
- Keep some flexibility in the research topic.

4.2.2 Recruiting PhD Students

Specifically when recruiting PhD students, check first that they meet the university requirements before even considering a follow-up with them. PhD student recruitment past the grades is more difficult to judge given they may not have had the opportunities to conduct projects and to publish papers or others. There is a lot of luck involved when recruiting a PhD student.

It can help to look for PhD students through your own network of previous colleagues or alumni offices. Postdoctoral researchers may be reached out through colleagues again, at conferences, or internally within the university. Word of mouth is more effective. Social media platforms can bring in a lot of applicants, but not necessarily the ones you are looking for. Watch out also for spontaneous applicants. They may not come at the right time, but reaching out to them later can be an option. Hence, storing out your contacts can be a good idea.

Recruiting PhD students means that you should be able to provide a scholarship to the student though a grant, or that they are awarded a scholarship. There may be also some other international scholarship available for certain nationalities, or scholarship with industry sponsors. Studying the range of scholarships available to tap into is helpful to redirect the students accordingly. However, some PhD students may also be able to self-fund their studies. If their CV is good and they are a good fit to your research, but there is no scholarship available for them, you can ask them if they are able to support themselves.

PERSONAL SHARING

In many universities, having a master's degree is a requirement for entering a PhD program, but in my university, it is not the case. Because my research is heavily experimental, I find PhD students having previously earned a master thesis to be more able to conduct lab work autonomously quickly, whereas PhD students with only a bachelor's degree need a lot more direct guidance to handle lab experiments on their own. For example, some students may not know how to effectively collect and store their data or even use equipment as simple as a balance. Training them means there must be a senior researcher available with them in the lab, which is quite challenging to do as a professor.

KEY IMPORTANT POINTS

Recruiting PhD students:

- Check they meet the university's requirement first.
- There is a lot of luck involved in the process.

4.2.3 Enrolling Short-Term Students

Enrolling undergraduate students for short-term projects or master students for their dissertation may be simpler as they likely come from the same university. It may also be compulsory for you to take on undergraduate students if they have to conduct a project to complete their degree. If this is a case, given the number of undergraduate students you will have to supervise, one suggestion is to first ensure there is another PhD student or researcher who can do the direct mentoring, and ask that person to conduct the interview with the undergraduate student. First come first serve approach can also be time saving, as it is very hard to assess the best fit of an undergraduate student to a project. Besides, it is the opportunity for them to explore something new, and it should be de facto assumed that they would not have or need previous experience. Master students doing a master thesis may also have various degrees of motivation depending on their aspirations. It is good to get to know about their future plans beforehand and let them know of your expectations.

Recruiting student assistants for specific research projects can be somewhat hard since students also have to attend lectures and do their studies. Universities usually have various schemes for undergraduate students to carry out small research projects in parallel. Posting some broad and generic advertisement on undergraduate's student groups or projects platform can help. Word of mouth is also efficient to reach out to undergraduate students as they may not use the same social media platforms as professors do.

Recruiting foreign exchange students can also be an opportunity to get manpower to help PhD students or researchers. You may be directly approached by foreign students or may tap into the university's exchange programs. It is also difficult to predict how motivated the student will be and how much they will dedicate to the project, even if it is graded.

Overall, when welcoming undergraduate or other short-term students, it is highly recommended to ensure there is a researcher or PhD student that can mentor them directly. Because training a student takes time, and training a student that does not deliver can be frustrating, the mentor should be aware of this too.

PERSONAL SHARING

I was looking for an undergraduate student with business training to do a market research and business plan assessment for one of the technologies developed in the lab. Asking other undergraduate students currently in the group if they know anyone lead to find a student very quickly. Because they were friends, it made also their collaboration and communication easier as they were working on the same project, but on different aspects. In other instances, I leveraged my alumni network to enrol exchange students. The enrolling part was challenging due to the darkness of university's administration, but their presence and their work initiated a lot of enthusiasm and ideas among my own team, especially because they came with a slightly different background and experience.

KEY IMPORTANT POINTS

Recruiting short-term students:

- Tap into student's network.
- Make use of the university platforms.
- Ensure there is a researcher or PhD student able to be a mentor.

4.3 Supervising Students

This subchapter discusses how to efficiently manage a research group, from setting clear goals and expectations for each group member and communicating them effectively. How to guide PhD students in writing their PhD thesis as well as managing and helping students with personal difficulties or issues are also presented.

After staffing your research group with students and researchers, it is important to guide them effectively to achieve the desired outputs. This is particularly important if the student or researcher is funded by one of your grant: they should contribute to achieving the KPIs for that grant. Even if they are not attached to a grant, it is good to develop a strategy to guide them effectively to explore new areas of research. As each project is defined with specific scope and tasks, you can re-think about what concrete output can be expected from each student and researcher according to their competency, as well as how they fit within the entire group.

Supervising students and researchers requires you to find a balance between micromanaging and letting them go on their own. One approach to finding the right balance can be to try and see, as well as to ask them what their needs and preferred mode of supervision are. Without big surprise, the higher the experience of the student or the researcher, the more independent and autonomous they become. However, there are always exception and being able to adapt to each individual is needed. The amount of time and attention you can give to each team member also depend on the size of your group. When starting, it is advisable to start small and slowly grow, so that you have enough time to find what works best for you. Having a well-structured group structure, slightly hierarchical can also help with the supervision. This means that undergraduate students are mentored in the lab by postgraduate students who are themselves mentored by postdoctoral researchers.

4.3.1 Setting Goals and Expectations

Setting clear expectations and goals for each team member can also help them having a framework, especially for the undergraduates and PhD students which may still have to juggle courses with research work. Typically, the goal of a PhD student is to publish a certain number of papers and to write their thesis, the goal of a postdoctoral researcher might be to publish another number of papers in a preferred list of journal, to set up an equipment, to write a technical disclosure, etc. While setting goals is helpful, they can also be revisited later in the project.

Informally sharing what a good student should be in your views can also be a way to give them a model or an example of your expectations. As stated before, understanding the motivations of each member will help you guide them appropriately. With time and across cultures, the aims and motivations for students or researchers to join your group likely are different from your own motivations. It is possible to find out by asking them directly, discussing with colleagues or other students, or even through literature. There is an ample number of articles discussing mentoring and supervising PhD students, some of them are listed as additional references at the end of this chapter. Reading them with the students can help clarifying the expectations and making sure you align.

Setting clear goals and expectations at the onboarding of students and researchers is needed also to ensure that they finish their work before leaving. It is unavoidable that some data do not get published, but given that quitting can happen quickly and sometimes unexpectedly, ensuring the data will be collected and that the results will be shared before leaving can help with the frustration of not finishing a work. Some postdocs or students continuing in academia may be willing to publish their results later but when they go to other jobs, it is unlikely that they will have the time and drive to finalize their past publications. This means a lot of energy and efforts placed in the project might be a little bit wasted if the data are not regularly collected.

Although as an assistant professor on tenure track, your own agenda is quite high level, be reminded to have reasonable and ethical expectations for your staff and students. There are always examples of high-profile professors putting a lot of pressure on their team and having questionable working ethics. For example, some people may be conducting group meetings on weekends or requiring students to clock a certain number of hours. Overall, every postgraduate student and researcher works very hard. It is not uncommon that students and researchers prefer to work in the evening or at night, when the facilities are not over booked. As you do too, your students and researchers also need to relax. Transferring the stress and pressure you may have onto the students and researchers may be counter productive. Check also your university for any policy about working hours and observe to find the overall working culture to align with the standards.

As the head of the group, you serve as a guide if not a model for the working habits of your group. You can also educate yourself on the risk of burnout, depression amongst students. Different countries and cultures come with different pressures to succeed in their work. Having basic knowledge on how to handle first response to assist student in distress can unfortunately be quite useful. As the direct supervisor, it is possible to encounter students breaking down in your office and sharing difficult moments. Being aware of the supporting structures in place in the university is very useful in these situations. Managing a team of people means there is likely a whole range of personal and professional issues that will come in. Examples can be having a couple in your team and there are tensions between this couple; having improper behaviour between two members of the team; having a researcher breaking one university rule without your knowing, etc.

PERSONAL SHARING

I defined to myself what a student of my dream should be, which helped me set the expectations for the incoming students. Sharing expectations and goals are key, while it is also necessary to be able to adjust for each student. Usually, I also ask them to come up first with their own goals and timelines, then I share what I expect and how I think they should do. This exercise can help both of us readjust accordingly.

KEY IMPORTANT POINTS

Setting clear goals and expectations:

- Find an appropriate mode of supervision.
- Define what should be the key performance indicators for each project.
- Be reasonable and ethical.

4.3.2 Preparing Students to Their PhD Thesis

PhD students work towards their PhD dissertation and defence. The writing part of the PhD dissertation can be stressful and time-consuming given the amount of data gathered. One approach to the PhD dissertation writing is to structure the dissertation early on with the student, meaning to decide what should be in the chapters already after 1 or 2 years in the PhD.

A PhD structure is typically: Chater 1 (Introduction), Chapter 2 (Literature review), Chapter 3 (Project 1), Chapter 4 (Project 2), Chapter 5 (Project 3), Chapter 6 (Conclusions and future work). Following this typical thesis structure, it is possible to envision each Chapter 2–5 as independent research papers. Based on this, the student can work on each chapter separately and at the same time write the corresponding research paper. Introduction and conclusion can be written at the end, in order to ensure there is a tight cohesion throughout the thesis and between the different chapters.

Because in some universities, it is not allowed to just copy-paste the research papers to make the thesis, a little bit of rewriting can be necessary, essentially to avoid repetition of the methods and to better introduce each chapter. Having this approach places the emphasis on the publications instead of the thesis itself and may shorten the writing stage of the PhD dissertation to a couple of months instead.

PhD students may also be stuck during their research, or spend months working hard but not gathering meaningful data, which may be discouraging. Regularly checking with them through the regular meetings helps evaluate the progress and gently pushing in a more promising path. Suggesting them actively to attend some talks, conferences or workshops, can also be a means to unstuck and get new ideas. Overall, deciding from the start that there is one main project, and some side, back-up projects also permits to have a back-up plan if the main project does not turn out successfully.

PERSONAL SHARING

I find that for some PhD students it is challenging to write papers and to understand what is expected in terms of outcomes. However, because they have to write a report in mid-term (within the first 2 years of their PhD) in order to continue, it is at this stage that the structure of their PhD dissertation and which scientific papers they will write becomes clearer. Often, as professors, we need the publications and having a student writing a dissertation without the paper in mind means that likely it will not be published. Hence, I usually tell the students to write for a journal paper first, then to place it in their dissertation, and only the last chapter they are still working on they can write in the dissertation, and later I can help them transform it into a journal paper.

KEY IMPORTANT POINTS

Preparing students to during their PhD thesis:

- Plan the chapters early.
- Have a main project and side back-up projects.

4.3.3 Helping Students with Issues

Across the 3 or 4 years of the PhD studies, life events can happen, and some students may encounter personal difficulties that slows them down or prevent them for completing their thesis. What can be done to help them?

The first thing to do is obviously to show empathy and to try to put yourself in their shoes. Ideally, letting them explain what the challenges are that they are going through will enable to determine if it is thesis-related or not. Words of encouragements are always helpful, as well as showing understanding and allowing the student to take a break or a rest. In general, most PhD students work very hard and want to do a good job.

PhD-related struggles can be mitigated by sitting down and spending more time 'micromanaging' the work. It may be possible to assign some parts of the work to other students or researchers if these are the issue, for example to have others help conduct an experiment or collect an important set of data. If is not possible, it may be appropriate to tweak the direction of the research towards something that the student is comfortable with.

Other PhD-related struggles may be eased by providing more support, under the form of direct guidance. It may be possible to ask another senior student or researcher to provide more mentoring. It could also be possible to ask the student to seek help and advice from other faculty, for example if they have a thesis advising committee.

In some cases, the struggles are not PhD-related and taking a rest does not solve it. The student can then be referred to the special help centre at the university if necessary. If the thesis is really stuck, and the student is not feeling well, it would be good propose an exit strategy. Such exit strategy could be to finish one experiment/chapter and submit, if the level is almost there. Then to tell the student to take a longer break and to continue later, if this is a possible option. More drastic exit strategy is to quit the thesis and to find a more suitable position elsewhere. It is possible to help the student find another position through connections or the university may have schemes to transpose a PhD thesis into a master thesis.

One important point to pay attention to is that PhD struggles become very quickly PI struggles and everybody's struggles. Addressing issues early and effectively is key. However, letting the student decide what is good for them is also important.

PERSONAL SHARING

I had one PhD student who had a lot of personal issues, and I feel I had tried everything. First, I was expecting these issues to surface at some point as they had told me in advance that they had a special condition, despite initial efforts to avoid those issues. When issues arrived, I felt I tried everything I could: giving time, being patient, referring them to other students and researchers, even faculty. In the end it became too painful for the student, and nothing I could do, so we found the best exit strategy was to transfer to a masters' degree instead of a PhD degree. I had another student with a little of struggles too, but giving more close tips on what I want in terms of papers and allowing them to rest was sufficient. I would say that having them communicate their struggles and their needs openly is quite helpful for us as a PI since else we may just not be aware of those issues. I ask my team members to be open about these in my first meetings with them so they know there is a safe communication channel.

KEY IMPORTANT POINTS

Helping students with issues:

- Show empathy. Let them explain and decide too.
- Find alternative sources of support.
- Have an exit option.

4.4 Practical Tips and Procedures

This subchapter completes the topic of supervising students and researchers by providing practical tips for the organisation of the research group as well as for supervising different categories of students and researchers. Examples of these practical tips are distributing tasks, sharing procedures and rules on an intranet, and making students and researchers responsible for their own work. Setting a structure such as having a group name, an address, a website, a social media group, etc. can be quite helpful.

4.4.1 Group Name

I would recommend having a group name. This is quite helpful for branding yourself as well as to instil a sense of belonging to the team members. However, it might be tricky to find a group name that can represent your research for your entire career. Acronyms can have different meanings in different languages. Also, your colleagues may remember firstly your personal name. Look at what others are doing for their research group names to find inspiration and what is suitable for you. Changing the group name is of course possible but defies the purpose of creating a brand. One solution to this issue could be to use your own name. At least, it is unlikely that it will change.

Along with the name, having a logo can be nice: a simple logo can be easily recognisable and placed on PPT presentations, as image on social media, on recommendation letters or even personal name cards. You can decide to have a logo or not.

These details specific to your research group can be added in the email signature to reach a wider audience and make people familiar with it.

PERSONAL SHARING

I had initially chosen the acronym 'LDMC' for my research group, to represent Laboratory for Dense and Multifunctional Composites. That name was quite fun, and my first small group of students even came up with poses that represented each letter from the 'LDMC' acronym. But when I came closer to preparing my tenure package, I have changed it as I felt I am not researching only on dense and multifunctional composites. I found that something broader would be more suitable and allow me more freedom to explore any research topic in the future. Indeed, if there is a very specific brand, then it is hard to deviate from the brand. I also googled 'LDMC' but my group did not show up first, whereas I am pretty sure nobody else in the world has my name and people usually remember it as it is unusual.

KEY IMPORTANT POINTS

Group name:

- Having a group name and a logo can reinforce cohesion, team bonding, and visibility.

4.4.2 Group Website

Having a website or webpage is also recommended. Websites can feature a lot of useful information for your team members as well as for people outside the group. Having information about the group is important for recruitment. Having a website means there is a web address that can be shared as the email signature of each team member, thereby representing the lab and increasing visibility.

The website can comprise the following tabs:

- Home page: with recent news from the lab, such as a new member, a new publication or conference presentation. This informs each team member if they have missed anything and is a good interface with the public.
- Research: a presentation of the research conducted in the group in simple and concise terms. This can clarify to the team members what the research is about in its larger scope as well as inform prospective students or researchers.
- People: a list of the team members with some very basic and non-private information such as last degree or research topic. This helps keeping track of the group members. This may be particularly helpful for bigger groups as well as when group members are not meeting regularly. It may be recommended to add pictures when there is a large number of students.

- Publications: a list of the publications with all authors and journals, conference publications, patent applications. Academics take pride in their publications and acknowledging them on the website can be an indirect way to show and advertise their work.
- Outreach: a page to list the conference participations and other activities.
- Gallery: to showcase the group pictures and interesting pictures from the research.
- Job opening: you may have a page to post job advertisements.
- Teaching: you may also have a page to show your teaching activities. This can also include workshops or other teaching activities.

Having the work of your team members posted on the internet may be a simple gesture of showing how you value their work beyond the published papers. It also comes very handy to track records and to communicate with the public.

The website can also comprise an intranet accessible only to current team members. It is possible to allocate to a student the task to update or manage the intranet and to allow uploading new files that are useful. It is advisable to inform that the content shared should not be communicated outside the research group as there may be more sensitive information such as research recipes and personal phone numbers. No confidential document should be shared on the intranet of course.

The intranet content can include:

- General resources with templates for reports and presentations, general guidelines for the group, forms to access various laboratories or list of equipment, phone directory.
- Meetings with the planning of the group meeting dates and presentations, some information about when people are on leave, sharing of presentations slides
- Social with pictures taken at occasions
- Writing documentation with documents related to writing tips, poster preparation, reports
- Experimental and analysis protocols with chemical list, risk assessments, shopping list, equipment manuals, software tips
- Tutorials and lectures with links to slides or video recordings to relevant courses or presentations.

PERSONAL SHARING

Because I have to host quite a few undergraduate students each year, who work more closely to the PhD students than me and that I therefore meet about 1–2 times a year, having their names and pictures posted on the website is a great way to keep track of who is who and to recognize them. I am also hoping that they can make connection internally using the website resources. I got this idea of the website looking at other faculty's websites.

KEY IMPORTANT POINTS

Group website:

- Creating a group website is great for visibility as well as organisation.
- The website can also comprise an intranet for sharing documents with the team.

4.4.3 Internal Social Media Groups and Mailing Lists

Internal social media groups such as WhatsApp or MSTeams groups are great platforms to create internal support and cohesion within the group. You can have multiple groups, if necessary, for example one that includes current and past members, and another one managed by the current team members only for practical access to documents, equipment, or other topics.

Such internal groups can be used to congratulate students and researchers about their achievements, advertising open positions, sharing interesting lectures. It is always rewarding to be congratulated by the other and past team members. The groups can also include laboratory managers or external collaborators who work closely with the group. Keeping in touch with the alumni can also be a nice way to create a community.

These groups will make the communication between the team members smoother and create some cohesion within the group. This might be particularly useful when all the members are joining and need some support and guidance, or when the research is carried out at different sites. Digital information can also be updated regularly, which means that nothing is carved in stone. Phone numbers or group members can be added or removed, the website updated. Creating a special mailing list may also be convenient to directly forward relevant emails and important information regarding undergraduate reports, for example.

PERSONAL SHARING

Social media are great tools but still need some guidelines and trainings. While I have personal accounts on Linkedin, Bluesky, even Instagram, I also created Whatsapp groups that include my group members. One group is for everybody, including the alumni, and another group is for the members only. Initially, I was in both groups that I created. However, I found that the students did not really use it much. I was not sure if it is because it was too early in the research group age, the students just joined, they do not know if they should use them or not, or maybe they are just useless, or maybe they are 'scared' of me. Later, once I had enough group members, I exited one of the group and I think now they are using it more, for example for equipment sharing or organising events. For the other Whatsapp group I have, I prefer to clarify first what the group is for and what kind of information should be shared. Typically, I inform the students or researchers that this is to share general announcements only. I mostly use it to post new students joining, new papers, new highlights. Because our Whatsapp group collects mobile numbers, I also need to remind all students to not share the phone numbers outside the group and to not use the phone numbers for inappropriate communications (something that has happened before).

KEY IMPORTANT POINTS

Internal social media groups and mailing lists:

- Having some closed internal social media groups and mailing lists can be very helpful for group cohesion and bonding, or for sharing information relevant to the group only.

4.4.4 One-to-one meetings

Although not everybody may practice this, scheduling regular meetings can help students and researchers get organised as well. Regularity makes the environment more secure, with some flexibility of course. It also ensures that all team members get a chance to meet and talk to you, the professor, and get feedback and guidance. Blocking the calendar with regular slots for meetings avoids difficulties in scheduling meetings. The regularity of course can vary depending on the needs. Every 2 weeks for starting PhD students or every month for PhD students who have defined their research well, and for postdoctoral researchers, can work well. Because the meetings are regular, when something happens, such as some personal issue with the student or researcher, it will become very evident, as they will likely repeatedly cancel the meetings.

It may be a good idea to instruct in advance how you plan to conduct these one-to-one meetings. For example, if you expect a presentation with a PPT or other forms of data sharing. Some personal tip is to ask the students and researchers to write a short and concise report beforehand and to send it to you a couple of days in advance. This practice gives them time to sit and think about their results and their progress. Browsing through it briefly in advance as the professor enables you to digest the content and be reminded about the details of the project. This is particularly useful when your group starts getting bigger as it avoids having half of the meeting trying to remember what their project is all about. Because there is a clear agenda, the meeting can be conducted more effectively. Regular reports can also be a form of data keeping, which facilitates the data transfer when the student or staff is leaving the group. For PhD students, it is also a good way to witness and keep track of their progress.

An example of suggested structure for a regular report to send before a one-to-one meeting can be:

- What you did and what you have achieved:
- What has failed or where you had difficulties:
- Your plan for next steps:
- Other things:

Other types of one-to-one meetings could be to either you asking the student or researcher for a meeting to get an update or waiting for them to prompt you. This can be effective to help them being clear and efficient during a short meeting but may give you less ability to follow the project closely. Of course, you should find your own style.

PERSONAL SHARING

With the reports being sent a few days before each meeting, it becomes very evident when something goes wrong and there is an issue with the student: usually the reports disappear. It also becomes very evident when the student is blooming: the report gets longer and the student has more to share. I also find these reports great for record keeping and makes it easier to find the data when the student or researcher has left or for preparing conference presentations or other meetings. Some flexibility are of course also needed to allow adaptation to the current research progress. I find it is also a good practice for the student to write and present their data effectively.

KEY IMPORTANT POINTS

One-to-one meetings:

- Find your way to have regular updates from your students and researchers, such as one-to-one meetings.
- Inform them how you want these meetings to be conducted.

4.4.5 Group Meetings

Regular group meetings for everyone to attend are also good to have an overview of the research. Presentations for team members may be every 6 months if the group meetings are about every 2 weeks. The group meetings can also for example include a research presentation and a journal club presentation. The research presentations can be a good training for students to give 15-minute conference-like presentations summarizing their work. It may help them to learn how to craft a story and how to speak clearly and concisely. The journal club presentations may train the students in reflecting on the journal papers as well as sharing interesting findings.

An example of journal club presentation template may include:

1. The choice of the paper
   - Scientific merit: why did you choose that paper? Why do you think it has something special?
   - Presentation and literacy: what is the target audience? Is it well-written and how are the displays? What is the structure of the paper?
2. Introduction
   - What is the topic and why it is interesting?
   - What is the problem they are trying to solve?
   - How did other researchers try to solve that problem?
   - Why are current solutions not satisfactory and what could they do to improve?
   - What is their solution?
3. Materials and methods
   - What is the material, or the system studied?
   - What tools do they use to prepare/characterize/measure...?
4. Results and discussion
   - What did they obtain?
   - Was it expected?
   - How does this compare with what other people have done/ the literature?

5. Outlook
   - What was novel in their study? What did we learn from them?
   - What do they plan to do next? What else could we do?
   - What is your personal general opinion about the paper?

The group meetings are also a safe place for students and researchers to ask clarifications, share ideas and discuss the progress. As a professor, it is also a good practice at the start to learn how to chair a meeting.

PERSONAL SHARING

I also use group meetings for students to rehearse their defence or other important presentations. It is an opportunity for everybody to ask questions. Our group meetings can also count as seminar attendance for undergraduate students doing their final year project in the group. Pointing out very good presentation can inspire others to improve their presentation skills too. Once a year I share about the overall group research vision, accomplishments and future plans during the group meeting.

KEY IMPORTANT POINTS

Group meetings:

- There are many ways to organise group meetings, find the one that matches your style.
- Teach the students about what is expected in these meetings.

4.4.6 Communication Tips

Efficient communication does not come always naturally, especially in a very international environment. If there is possibility to use your native language, other than English, with some team members, it may help communicating. In this case, one needs to be careful to turn to English when there are non-native speakers of the language joining the team. Given that most conferences, journal papers and other forms of communication are usually in English, it may be good to practice English whenever possible.

In an international group where English is not the mother tongue to anyone, it may take some time to adapt to each other's accents. One way to ensure that the other person is following the conversation (and not only saying yes to please you) is either to ask them directly if they understood, ask them to repeat, or to take notes with them or to let them take your notes to check them afterwards. It is quite common that foreign students are more familiar with written forms of a foreign language than oral.

Similar to language issue, leading styles are very personal and may take time to find and to adapt too. If your style is to 'suggest' instead of giving orders, make sure that everybody gets it that they should take your suggestions seriously. At times, it may be needed to be crystal clear.

In addition, it may happen that the hierarchical difference of being a professor leads some team member to avoid sharing issues or difficulties. If necessary, popping into the laboratory to check how things are going and offering help can be very welcome. Likely, as a PI now mostly sitting in an office, it is always enjoyable to go back to the laboratory or the experimental bench. Some students may need to be a little bit taken by the hand initially to get started. Adjust accordingly.

It may also be necessary at some point to let your team members know what the boundary between professional supervision are and more personal feedbacks. This is because, especially for PhD students, the boundary between outside life and studies are even more difficult to draw than for working professionals. Let them know how much it is appropriate for you.

PERSONAL SHARING

I have a small note next to my desktop to remind myself to be cheerful and enthusiastic when I get a visit from a student or researcher sharing about some technical or other issue. Reading their reports earlier is helpful to have a solution or approach to address challenges, even at days where you do not want to hear about problems. However, I rarely take impromptu meetings when students just show up at your office and disturb you in the middle of something.

KEY IMPORTANT POINTS

Communication tips:

- Efficient and effective communication is a must. Take some time to find an appropriate and effective way to communicate.

4.4.7 Survival Guide

Students and new team members always come with the same queries. It can be helpful to come up with a 'Survival guide' to help them preparing before joining the research team. Team members who are leaving can help update the survival guide regularly. This can save your time in the future.

Examples of inputs that the survival guide can include:

- A welcome message with a general introduction of the group.
- Practical details about the organisation of the group such as location, access, affiliations, etc.
- What to do upon arrival such as what formalities to do after entering the country, at the university, with basic recommendations in terms of housing, banking, phone numbers, health services, etc.
- Some guidance about how the university works such as booking of meeting rooms, what to wear, how to apply for holidays, what recreational and dining opportunities there are, etc.
- How your group operates, for example how to book an equipment, how to print, how to have access to the laboratory, how to order consumables, etc.
- Indications about meetings and project organisation, such as your expectations, some reminders about ethics, what to do when wanting to write a paper or join a conference, details about data management, etc.
- More specific details for graduating students about the PhD timeline and expectations, how to submit the thesis, other requirements, etc.
- Some tips about work-life balance and stress management such as information about impostor syndrome, where to find support, where to find inspiration, etc.
- Some suggestions for their next career steps, how to keep in touch, etc.

PERSONAL SHARING

I find the survival guide particularly helpful to address certain questions from visiting international students, which can span from 'how to dress in the laboratories in Singapore' to 'what kind of scholarship I can apply to'. To keep the survival guide updated, I ask everybody who leaves or newcomers to go through it and update the sections that are not relevant anymore.

KEY IMPORTANT POINTS

Survival guide:

- Creating an on-board document to share some tips, rules, or practices in the group can help ensure everybody is on the same page when joining.

4.4.8 Welcome Slides

Along with the survival guide, it may be timesaving to have a regular on-boarding process with sharing the critical information orally in a meeting. For example, having a welcoming list or welcoming slides to go through with any new team member joining can clarify things at least once. Sometimes, it is good to get a reminder about obvious things like no plagiarism, especially if the team member is new to the university. The list or the slides can be uploaded on the intranet or shared with the team member if they need to consult them later. It also gives them the opportunity to ask directly some questions if anything needs to be clarified.

Examples of welcoming slides may include:

- A general introduction to the laboratory
- Presentation of the facilities and equipment
- What are the resources available
- How the lab is organised
- About publishing, patenting, and presenting at conferences
- Rules about data management (usually every institution has some rules to follow)
- Rules about human management (what to do if there are some internal issues)
- What are the expectations
- Reminder about ethics and proper behaviour
- What are the next pragmatic steps to get started

PERSONAL SHARING

I tend to forget to remind the incoming students and researchers about basic, fundamental expectations and rules and use these welcome slides to point out the importance of ethics and appropriate behaviour. I also use this opportunity to express the expectations within the group, such as collaboration and teamwork. Having the slides available to them later allows them to check it in case. Setting this open communication channel first is key as many students are shy or intimidated or just do not know what is expected from them.

KEY IMPORTANT POINTS

Welcome slides:

- While the students may consult the survival guide on their own, having a welcome slide deck can help highlight key important points.

4.5 Conclusions

Supervising students and researchers is key to accomplish your research agenda. You should recruit students and researchers strategically to meet the requirements of your project as well as to fit with your views for your research group. Setting clear goals and expectations for each individual project is important for good delivery. You can provide the scaffold for your team member to work on by creating a cohesive group with efficient and clear communication standard, regular meetings, and shared documents.

Additional References

Lee, C. Dennis, P. Campbell, Nature's guide for mentors, Nature 447 (2007), 791.

Graduate student-supervisor expectations, The university of British Columbia. 11 April 2014, https://gradstudies.ok.ubc.ca/academics/student-supervisor-expectations/, Accessed 30 July 2019.

J. Handelsman, C. Pfund, S.M. Lauffer, C.M. Pribbenow, Entering mentoring, a seminar to train a new generation of scientists, The Wisconsin program for Scientific teaching (2005). www.hhmi.org/grants/pdf/labmanagement/entering_mentoring.pdf, Accessed 27 March 2018.

Ryerson University, Graduate supervision guidelines. Aug 2018 //efaidnbmnnnibpcajpcglclefindmkaj/https://www.torontomu.ca/content/dam/graduate/current-students/policies/Graduate-Supervision-Guidelines-Aug18.pdf, Accessed 30 July 2019.

University of Copenhagen, Department of Science Education, PhD supervision: lead the way. https://www.ind.ku.dk/english/projects/phdsupervision/, Accessed 30 July 2019.

University of Groningen, Supervising PhD students: 'the 10+ commandments'. 25 Aug 2014. https://www.rug.nl/research/behavioural-cognitive-neurosciences/education/phd/supervising-phd-students_-the-10_-commandments?lang=en, Accessed 24 Dec 2017.

The Dos and Don'ts of PhD supervision. 7 Sept 2017, https://forgingconnectionsblog.wordpress.com/2017/09/07/the-dos-and-donts-of-phd-supervision/, Accessed 30 July 2019.

The PhD-Doctor: What (Not) to expect from your supervisor, 28 April 2023, by Herman Lilieveldt, Science Careers. https://doi.org/10.1126/science.abn2937, Accessed 30 Oct 2023.

5

Writing Grants

5.1 What Funding Do You Need?

There are many diverse types of grants and funds you can apply to. It is key that you select the ones you need and apply to them strategically.

5.1.1 Types of Grants

While there may be some small sources of funding allocated yearly to you to purchase small consumables and materials for your research group, it is likely not sufficient to conduct larger research projects. Applying for grants will not only provide you with the funds necessary to purchase the materials and consumables needed by your researchers and students, but also to get funds to hire researchers, PhD students, equipment, to pay for your travels and for the open-access publication fees. Most institutions rely on professors to obtain funding through grants for them to conduct their research.

There are various types of grants that you can consider applying to: internal grants which may be within your own department, school, or at the university level. These are likely to be of smaller amounts but may be less competitive; and external grants which are competitive at the nation-wide or international-wide level. External grants may therefore be individual grants, collaborative grants, fellowships or awards, or industrial grants. They all vary in their requirements and possible budgets.

How do you identify grants that are available to you to apply and which ones to choose? Very likely you will be informed by your university about the grants you can apply to. Asking colleagues or searching online can provide you with some ideas of suitable grants. As a new faculty member, it can be assumed that you will start with a start-up grant, which is a pool of money provided to you to start your research. From your preliminary results and research objectives, you should be able to identify what you need for your research and the type of research you intend to do: fundamental, applied, industrial, etc. This will give you the Technology Readiness Level (TRL) you stand at and which TRL you plan to go to, which may help scope down the grants suitable for your research. For example, industrial grants are likely to be for higher TRL.

DOI: 10.1201/9781003635116-5

As an early-career professor, you may also be eligible to apply to specific early-career grants and awards which may be less competitive that other general grants. Seed funding grants are also helpful to start new projects, with the aim to apply to larger amount of funding later.

PERSONAL SHARING

Grants obtained might be an important performance assessment at your university. Being able to get funds for your own research means you are able to be independent and to have leadership. Assistant professors on tenure track are more 'hungry' for grants and achieving high impacts which might put them in favourable positions for many grants as compared to tenured professors. Grants specific to early careers can be a great avenue of securing funds to start with. I also started with smaller internal grants that enabled me to gather preliminary data that justified larger grant proposals. Initially, I also applied to early-career grants as there was a smaller pool of applicants as compared to other grants. If possible, I prefer the 5-year long grants as they allow the development of a more substantial body of research, but shorter grants are also sometimes just necessary to get started on a new project and to demonstrate that you have expertise in a new area. For example, as most of my work is experimental, it is hard for me to propose a research project on simulations if I cannot show previous work on simulations, even if I know I can do it.

KEY IMPORTANT POINTS

Types of grants:

- Get an overview of the grants you could apply to.
- Identify which TRL you are and what kind of grants would fit.
- Consider applying for grants specific to early careers.

5.1.2 When to Apply

You want to ensure that there is always some on-going funding for your research group so that you have no interruption in your research work and publications. This requires a staggering of the grants, i.e. not having them all starting and ending at the same time.

You also want to have as much money as possible to do the research, but you also want to give yourself a reasonable amount of work and a reasonable number of students and researchers to manage. This means that while you can try applying to all possible and imaginable grant calls, it may be more thoughtful to apply to only a few that you have carefully selected. Else, in the scenario of all grants being awarded, you might not be able to cope.

You also want to have a few, but large projects. Since it takes a lot of energy and effort to write a convincing and competitive research proposal, larger grants, with a longer duration, will allow you more time and flexibility for doing your research. A 5-year grant is better than a 2-year grant.

Writing competitive grant proposals also takes time and you do not want to 'waste' all your great ideas directly. Indeed, very often, unsuccessful grant proposals cannot be resubmitted unless a significant change to the proposal has been made.

To find out when applying, you can first look at the grant requirements. If it is an early career type of grant, fellowship, or award, you can see how it may fit best within your career plan. You can also look at how many times you may apply if it is not successful the first time. For other types of grants, you can see in your research timeline when it would be better to apply as you may want to get preliminary data first, to support your proposal. The grant guidelines will also tell you the frequency of this grant and the deadlines.

PERSONAL SHARING

While I initially applied mostly to early career kind of grants, I soon looked into general grant calls. Drafting proposals, even if unsuccessful, gave me ideas for new projects and I could 'recycle' part of the grant text into research papers. Being invited to join other researchers on their grant calls early was an eye-opening to see how people manage collaborative grants. When looking at grant calls, which are sent mostly through the university channels, I first look if one of my existing ideas could fit into the grant call. I have a few ideas of projects I want to explore, if the grant calls feels the right fit, I would go directly and write the proposal. Usually those ideas have been matured over the years, and may also be from rejected proposals. I rarely do the other way round, as in: there is a grant call and I try to see if some of my work can fit in. I tried it once and put a lot of efforts into it. It was a food-related proposal, collaboration with overseas partners which I had to found. The process was enjoyable and meeting these new colleagues enriching, but I realized that the agency and the type of proposal they are expecting were new to me, which is likely why it did not work out. We crafted the project specifically for that grant call. On the plus side, now I have learned and could try again more successfully in the future.

KEY IMPORTANT POINTS

When to apply:

- Stagger applications to have always on-going projects.
- Aim for larger and longer projects preferentially.
- Check the frequency of calls and their deadlines.

5.1.3 Collaborative Grants

As an assistant professor on tenure-track, you have to establish yourself as an independent Principal Investigator (PI). It is a good consideration to try getting grants as single PI or at least as a lead PI to demonstrate your leadership. When starting at a new university, being single PI may also be easier as it takes time to find trustful collaborators. In some cases, however, it may be helpful to submit as lead PI with a senior well-accomplished professor as co-PI or collaborator, as they can provide you with guidance, although it should not be a previous supervisor.

If you have the chance to get invited to participate in a grant call as a co-investigator (co-I) or collaborator, you should of course also consider it. Likely, it will require less effort from you as compared to being the lead-PI. You will still be expected to contribute to the writing of the proposal, but not to coordinating it. It is a great opportunity to see how other people do it and to learn from more experienced PIs. However, you need to pay attention to how much budget can be allocated to you and if it can be allocated to you. Grants as co-I may have a lesser impact for your tenure application but can give you multiple opportunities to learn, enter a new research field, find collaborators, and become a good team player. It may be an opportunity for you to showcase multidisciplinary interests too.

Finding collaborators and co-Is to work with you can take some time and demand a little bit of effort and getting out of your comfort zone. You should be aware that being a co-I or collaborator may make the grant less of a principal activity for them and therefore may require more efforts from your side to 'push' them. You can tap into your current networks to find the people you need on your team. Another way can be to reach out to some potential collaborators 'out of the blue'. Most of the times, agencies putting grant calls that require multiple PIs will make some efforts of match-making between interested professors or researchers. This is the case, for example, for joint international grants between countries or between research centres.

PERSONAL SHARING

Usually, I prefer to ask colleagues I know well and am comfortable with to be my collaborators, or to suggest to me some people to join. Sometimes, I can find a good match with people at conferences or when visiting other universities. However, I have recently been reaching out of the blue to some collaborators for joint applications. To find these people that had the expertise I was looking for, I asked ChatGPT to give me a list of names. Then, I researched their profiles on the internet to confirm they matched what I needed. To approach them I wrote a simple email explaining who I am, what I would like from them, and why I think they are perfect for it. I asked to have a call to share more details. Usually if there is no interest or not a good fit, they would either not reply or say thank you only and it will be pretty clear. But so far, I found ChatGPT to be quite good at providing the right people. I used this approach for finding speakers for conferences as well, when I ran out of connections and this is also efficient. While none of the projects got funded, I think it was really nice to start the discussion and will definitively keep my eyes open for the next calls.

KEY IMPORTANT POINTS

Collaborative grants:

- Think strategically if you need to be PI or Co-I.
- Choose your Co-Is or collaborators strategically too.
- Reach out to get what you need.

5.2 Preparing the Proposal

After identifying your team for writing the proposal, the research project idea and which grant to apply to, it is time to start preparing the proposal, writing the research plan and aggregating the supporting documents.

5.2.1 Getting Started

Writing a grant proposal takes time. You need to clarify the idea, identify the impacts, the significance of the research, craft the budget, review the literature to stress the existing gaps and how you will address them in your project, proofread everything many times, etc. It is strongly advised to start early, at least a couple of months before the deadline, if not more, if the proposal involves several people, which will require extra coordination.

Before starting any writing, ensure you review the requirements and the deadlines of the grant call carefully. There may be documents that take some time to obtain, such as supporting letters from external parties, or you may need to request internal approvals from your research unit or university. You may also need to check what the procedure is for applying to grants and if there is a unit at the university that may review the proposal and provide some feedback. If there is any feedback, it is always welcome.

One other aspect that requires careful consideration is the grant portal. If you have to submit your proposal through a dedicated portal, requesting access may be necessary and may take some time too. Getting familiar with the grant portal is a good idea to avoid last minute hurdles. Most grant portals encounter some hiccups when too many people apply at the same time, or there are some blackout periods when the portal is getting updated. Sometimes, the information that you need to fill in the portal may not be written in the grant call documents. For example, inputting the ORCID numbers of each team member may be required, but you were not informed of this earlier. Hence, once more, applying early is a good idea.

Before getting started with the specific writing of the research proposal, identify the key novelty of the proposed research, its impacts and significance. Drafting a diagram on how to pursue the research with the inputs, methods, and outputs will help craft the core of the proposal. Having a schematic and a crystal clear idea of the research will also facilitate communication with your team if you are submitting with a team of other PIs. It will also make it easier for them to contribute and to identify existing preliminary data, published or unpublished, to support the proposal. You can coordinate how to start writing, for example, using a shared Google document, and decide who is writing what, if this is a collaborative effort. Make sure also to arrange regular weekly meetings to set up the collaboration if needed, and to share how to communicate (phone number, for example). Confirm everybody's availability too.

PERSONAL SHARING

When leading a project with several other members, it is sometimes difficult to make everybody happy. When the deadlines are quite short (some grants require you to submit within one month!), it may be necessary for you to give 'order' to your colleagues about what they need to do. Try to do this is in a collaborative manner, but sending multiple reminders, chasing people, and being extra clear about what you want and when will facilitate their work too. If you are a co-I, you will appreciate the clear directions. Balancing the needs of all researchers to fit in the budget of the grant may also be challenging. I find a good way is to allocate how much of the funds will be for each co-I, and then they can decide how to spend it. Again, so far professional communication always worked.

KEY IMPORTANT POINTS

Getting started:

- Be able to spell out the general idea concisely.
- Start as early as possible.
- Make use of shared documents.
- Be prepared to a little of rush/stress.

5.2.2 The Research Proposal

A research proposal contains several elements. You should follow the template or instructions provided by the funding agency. Make sure to fill up every section in the template and to leave nothing uncertain. Make sure to follow the format and page or word limit.

When preparing the content, remember that the reviewers are humans too: they get tired and/or excited. Use bold font, underlined text, or italics to highlight important keywords and make it easier for them to catch the most important elements. Make use of keywords to make your proposal attractive, while making sure to balance or moderate the use of 'buzzwords'.

Identify clearly what are the aims of the proposals, the hypothesis or the need you are addressing, what is the approach, and what are the expected outcomes. One suggestion is to write sentences such as: 'This project aims at …' to guide the reviewer to what is the general aim of the project. Specify the smaller aims clearly as well and introduce them clearly, for example by using the following structure:

'The specific aims include:

- Aim 1: …
- Aim 2: …
- Aim 3: …'

In the same way, you can underline the novelty by making statements like 'The novelty of the proposed research spans several aspects:

- Novelty 1: …
- Novelty 2: …
- Novelty 3: …'

For each of your aims, define the specific Work Package (WP) indicating the preliminary data, the approach, and the intended outcome.

Support your proposal with graphics and figures. A graphical abstract can make the proposal appealing and easy to understand and to be followed for the reviewers. Preliminary data and methodology or system studied can also be presented in a more concise way using a figure. Indeed, a figure is worth a thousand words.

Make sure to stress how the research is feasible by adding the resources and expertise available, while also pointing out the challenges and the broader impacts if the research is successful. It is usually recommended to balance risk and impact. A proposal that is too novel may seem less credible and less feasible, even if the impact is very high. A proposal that is too incremental and not novel enough may, on the contrary, not be very appealing to the reviewers. Adding some benchmarks if you try to develop a new technology or product will help visualize what are the benefits of such new technology or product.

Provide references to support your claims in the proposal. Depending on the length of the proposal and of the project, 10–30 references can be a good number.

PERSONAL SHARING

The best way to write your first grant proposal is to have a peek at other's grant proposals. For some very specific small grants, I had asked colleagues I knew were successful to see how they wrote. I found some grant proposals shared online on personal blogs and open access repositories, even for highly competitive research grants like for the European Commission. Afterwards, as I got some grants awarded, I used the same type of writing for my next grant proposals.

KEY IMPORTANT POINTS

The research proposal:

- Make sure to follow the instructions and the template.
- Guide the reviewers through the proposal by stating the 'aims', 'impacts', 'novelty', etc.
- Divide the research into work packages.
- Provide graphics and figures.

5.2.3 Other Documents in the Proposal Package

Besides the research proposal, you will need to provide other supporting documents and additional information. All documents that are required will be clearly indicated in the grant call documentation.

Milestones and a timeline of the proposed research are expected. A milestone can be an unquantifiable deliverable such as 'establishing the model to simulate the system' or 'obtaining the design for the system'. It is recommended to use verbs when defining milestones and to have from 1 to 2 milestones per year only, which makes 3–5 milestones for a 3-year grant.

Curriculum vitae of the investigators will be required. Having a short CV ready will help, although you will want to tailor it to fit and match the specific research proposal, for example by highlighting your papers in the area of the proposal or any other relevant skills. If this is a collaborative proposal, you can include your past collaborations; if you are proposing to disseminate the research at a conference, you can point out your past speeches or symposium organization roles, etc. If a template is provided, follow the template.

Letters of support or letters of intent from industrial partners may be advantageous to have, even if there is no commitment from the company and even if the research is more fundamental. These letters can put forward the potential impacts and the significance of your proposed research. Obtaining such letters requires to have industrial contacts and some preliminary discussions beforehand. Drafting the letters yourself to send to the industry is a good way to make it easy for them. Finding industrial contacts also takes time, and you can adopt the same strategies as when looking for other co-Is or collaborators.

You will need to provide a tentative budget which will account for the manpower, equipment, materials and consumables, travel, scholarships that you will need to conduct the research. The budget should, of course, be within the limits of the grant call and be convincing. Some would recommend crafting a budget close to the maximum limit of the grant, while being reasonable. Budgeting for equipment likely will require you to provide the quotations for the said equipment as a justification. Materials and consumables, maintenance, open access publication fees, local conferences, local travel, etc. should also be accounted for in the budget. Being able to name specific journals and conferences will help justify the need to budget for these items.

Key performance indicators (KPIs) should be provided. These include the number of publications, talks, students graduated, prototypes, patents, etc. that the research will deliver. These KPIs will be compared to what has actually been achieved during the research, hence it is critical to be reasonable and not to overpromise.

Other documents may be needed, for example, if the grant will support only a few years of a PhD students, then an additional letter from another grant or the university to complement the stipend and scholarship of the student might be needed. You may also need to submit a declaration of other sources of funding currently held or applied for.

PERSONAL SHARING

For the KPIs, I usually put one paper per year, which is reasonable to attain and I will likely publish more if there are several researchers working on the projects. In my first grant, I put 'start up' as KPI which, while I would love to have it, was totally unreasonable given the lack of experience and knowledge I have about entrepreneurship. I use my first 'early career' grant to realize all these mistakes and be more realistic in the next ones.

KEY IMPORTANT POINTS

Other documents in the proposal package:

- Check what documents you need to submit along with the research proposal.
- Usually: milestones, timeline, CVs, budget, and letters of support will be needed.

5.3 Outcome & Further Administrative Matters

Congratulations, you have prepared you grant proposal and have submitted it. A few months later, you receive the outcome, which will determine your next actions. For successfully awarded grants, more administrative matters will come in as you start and conduct the research.

5.3.1 Outcome

It may take 6 months to a year, roughly, between the time you submit your proposal and the time you receive the outcome. Make sure to save all your files in separate folders and to keep track of the grants you have applied for.

The evaluation of grant proposals may take several steps. Some require just a file to be submitted, while others include one or several rounds of interviews or submissions. Each stage should come with an outcome and feedback.

If the final outcome is positive, congratulations! This is a big thing and you should celebrate it. You might still be asked to make some amendments to your proposal and to provide a more specific budget. It can happen that the budget awarded is smaller than the budget requested, which means some readjustments to your exact plans should be made. You will also be asked about the exact starting date, which you may be able to defer by a few months, allowing more time to advertise positions and find potential candidates.

If the final outcome is negative, do not take it personally. It will eventually happen, and likely multiple times. To avoid getting discouraged, you can consider that there are a number of other aspects than the quality of the proposal that came into play during the evaluation stage. For example, the number of proposals and the overall funds available, if the funding agency was looking for a special type of proposal, even if the funding agency had done previous discussion with specific groups of researchers... who knows.

Unsuccessful grant applications are never nice, but it also does not mean that the idea is not worth it. Maybe the core idea could be reused for another application or in another type of context. Resubmitting the same proposal is not a good idea, but rewriting for another grant call, on the same idea but presented slightly differently could be worth it. In addition, you should also receive feedback from the reviewers which could provide insights into why it was not funded and help preparing for another grant call.

Rejected submissions may also be invited to be resubmitted, provided significant modifications are made and the comments from the reviewers are addressed. Usually, proposals invited for resubmission have a higher chance to be successful, although they go through again a new round of evaluation.

PERSONAL SHARING

I noticed that while I wrote proposals that were not successfully awarded, I wrote them because I really wanted to do these projects. Somehow, I sometimes managed to do what I wrote in those proposals one way or another, for example, by getting students to work on it and by fitting it in other on-going projects. While obtaining the grant would have allowed to go deeper into the research and explore other aspects, the writing of the grant itself helped me in setting the steps to conduct that specific project which made it a lot faster to happen. In another case, the rejected grant ideas have been in the back of my head since and waited until recently to submit the proposal which got awarded. The initial proposal was too early as I could not show convincingly I could do it due to the lack of preliminary data in that specific area, and therefore could not convince that I would be able to do it. Waiting a few years and having papers using similar approach and method, now I could state in the proposal that I had the experience to conduct the research successfully. In another case, one of my grant was unsuccessful but invited for resubmission. I thought I would make a stronger case by bringing in a 'big name' in the area as a collaborator. The resubmission was successful (but I am not sure this addition was the reason).

KEY IMPORTANT POINTS

Outcome:

- Grant proposal evaluation may have several stages.
- Feedback should come along with the outcome.
- If unsuccessful, find ways to make use of your efforts.

5.3.2 Administrative Matters during a Project

During the research project, other administrative tasks will be needed. The first one will be to craft and edit the data management plan. This ensures that the researchers and students participating in the project have a clear understanding of the data collected and how to store it for future years. Typically, data should be collected for 10 years. You can check with your university on the policy for data management.

Yearly progress reports or final progress report will also very likely be requested by the funding agency to know what has been conducted during the research and what are the impacts. They may check on the progress milestones and ensure the KPIs stated in the proposal have been met. Any deviation from the proposal should be clearly justified. They will also check that the budget has been utilized, and it is good to use as much as possible the entire budget awarded. These reports may impact future funding by the same agencies if they are not satisfactory. Underspending will not be regarded favourably and may lead to you being awarded lesser budget in future years or the fund for the grant to decrease.

Budget variations, changes to the research plan, addition of collaborators or change of PIs, and grant extensions can be requested according to the grant guidelines. While all these are feasible, they will require some time to be approved and processed, so be aware of this and plan accordingly.

PERSONAL SHARING

I have made about 20 budget variations in my first research grant of 5 years. While it always takes some time to fill up the form, and to ask the administrative people to check it and get the approval of several people each time, I needed to do this because (1) I was inexperienced and my initial budget was a bit inadequate, (2) I wanted to use most of that budget, (3) some unplanned change in the research direction had to be taken into account (for example, COVID, or an equipment that we actually built). Then, my second big grant of 3 years, I made only 2 budget variations.

KEY IMPORTANT POINTS

Administrative matters during a project:

- Reports will be requested during and at the end of the project.
- Any deviation should be justified.
- It is ok to make grant variations, but it takes time and effort.

5.4 Conclusions

Mastering writing grant proposals and the management of projects is very important to position yourself as leader in your field and to become an independent PI. The first grant proposals you write can be your learning platform to improve. Managing the administrative paperwork is cumbersome but necessary. The faster you learn the easier it will be for you to navigate the funding landscape. Optimize your efforts and do not 'waste' any unsuccessful grant proposal. Explore various types of fundings and apply regularly.

Additional References

K. Powell, The best-kept secrets to winning grants, Nature 545 (2017), 399.

Writing competitive research grant proposals, Suggestions from Program managers of the ACS petroleum research fund, https://share.google/ZX9R7GUV1Av67Ul0p, Accessed 13/03/2018.

P.E. Bourne, L.M. Chalupa, Ten simple rules for getting grants, PLoS Comput Biol 2 (2006), e12.

B.M. Ishaq, C.H. Kumar, Crafting a convincing research proposal: a guide to writing successful funding applications, J Pharm Res 22 (2023), 8–20.

6

Writing Papers

6.1 The Need for Writing

Scholarly articles are one key metric currently used for evaluating the work of a professor. This subchapter discusses the different kinds of writings, such as research papers, reviews, and editorials, and how they can all be important for the job of a professor. Tips on how to publish strategically and efficiently are also shared.

6.1.1 Scholarly Publications

Scholarly publications or research papers are one key metric that is evaluated for assistant professors' promotion. You are expected to publish in journals with high impact factor. However, how you evaluate a journal to be of high quality may differ from the impact factor number. As mentioned in previous chapters, it is good practice to narrow down your research projects to scope what papers you want to write and identify to which kind of journals they could fit in. At the start of your independent career as an assistant professor, it may be difficult to publish high impact publications while you are setting up your lab and securing grants. Therefore, you can also target smaller types of journals first. Publishing in smaller journals while involving students and in particular undergraduate students can also bring value to your tenure package as a way to train future researchers.

When writing a journal paper, the order in the author list may matter. Different disciplines have different views on which order is the most important, and it may be good to check the expectations too. Similarly, whether conference and journal papers are the most highly valued depends on the field. Because you are expected to demonstrate your independence, you will need to publish as the main or corresponding author. It is also recommended to avoid publishing with your past supervisor. However, adding some collaborators can also be of value as it can help you publish into other fields than your own, or to show that your work may have multidisciplinary applications. Who is on the author list can also give a push sometimes to your

DOI: 10.1201/9781003635116-6

publication, for example by inviting a world expert in the field to collaborate with. Indeed, the author list, along with the paper title, is what the editors see first. Similar to the variety in funding sources, having a variety of publication types and journals is recommended. One aspect to pay attention to, however, is to avoid publishing in predatory journals. They may make your work seem less legitimate. Predatory journals also require fees to be paid to be published, so, it is not in your interest to spend valuable grant money to publish there.

You can decide how you want to publish based on your university's expectations and your own short- and long-term research aims. For preparing your tenure package, you will tie all your papers together to make a story. It is therefore important to write the papers that are meaningful and relevant to your research, and that you are proud of.

PERSONAL SHARING

Inviting a 'big name' can be helpful in publishing a paper. I have done this in the past. It allowed me to publish in a field slightly off my own field of expertise. However, you should clarify beforehand with the person how they will contribute to deserve their name on the paper. In my case, they read through the paper, provided feedback. It followed that we had many discussions about the topic and even thought of collaborations, etc. There are straightforward ethical guidelines on authorship based on the CreDiT statements. I am quite happy with that paper.

In another scenario, I had been invited to be named in one paper by a colleague at another university if I could help review. The authors also invited me to give a talk overseas, and we have more discussions, and the idea was to have a follow-up work together. Now I see they published the paper after I gave them some suggestions and did not put my name in. I think it is actually much better that way, because it was very odd previously. I will definitively never say yes again or be clearer on what I consider ethical authorship.

KEY IMPORTANT POINTS

Scholarly publications:

- Diversify your publication types and journals.
- Match your research aims.
- Be proud of your papers.

6.1.2 Publishing Strategies

Some common strategies to increase the number of research outputs include writing conference papers and then writing the full version of that paper. Although the impact of the conference paper may be small (if any at all), it can again be a nice bridge for students to learn paper writing and gain confidence. The literature review material used in writing a grant proposal can also be reused for a publication by expending the scope. Vice versa, a literature review you may want to write can then help you scope the research gaps to address in a future grant proposal.

When starting your research group, writing a literature review may, anyway, be a good idea to have a paper published when research data collection is in progress. To increase the research outputs when your team is small and just starting out, you may consider combining undergraduate students' projects together, for example by designing their research scope to be complementary. You may also consider publishing old postdoctoral data or even PhD data (upon agreement with you past supervisors). Although this may not be a strength to state your independence, it can nevertheless boost your citations count. Finally, you may also consider conducting your own research yourself and writing your own papers. This is possible for example if you can generate data without much time or equipment, such as using theoretical models, calculations, or developing conceptual frameworks.

To ensure continuous paper publishing and avoid publication gaps, one approach can be to always have one paper at one of the publishing stages, which is one where research is being conducted to gather data, one under drafting and one under peer review.

Different kinds of papers bring different values to your research portfolio. Literature reviews and perspectives can help you increase your number of citations and increase your h index, if this is still a metric used for your assessment. Journal and conference papers will help you build your expertise in a certain field and should be your major type of publication. Other types of publications can include research news or historical notes, which can help you showcase or practise writing for outreach and further impact. Writing editorials can show editorial role and impact in the scientific community too. Pure outreach papers in magazines can also be fun, impactful, and show the desire to communicate your research to a wider audience.

To make the most of your writing work, after the paper is published, you can share it on social media to help communicate it with your network and spread the word about it. Posting on social media can also help with increasing your citations and increasing your visibility and placing you as an expert in your field. At the same time, sharing on social media can also increase visibility to industry or potential future students.

Some new practices to showcase your research include making a video summary or another catchy visual that can easily be shared. If the paper has high impact, informing the university communication office may also help make a news release for broader impact. Some journals require that a graphical abstract is submitted with the manuscript. Graphical abstracts is a visual representation of the paper. They are ideal images to share on social media or to publish on your own website as they represent the work in one glimpse. Getting a cover image on a journal can also be a great occasion to make nice and catchy images that can be shared to highlight your work.

Finally, most funding agencies now require the papers to be available in open access, either their final published version or their accepted version. Hence you may want to upload the paper on open access platforms such as your university repository or other open access platforms. You can use Open policy finder (https://openpolicyfinder.jisc.ac.uk/) to understand the embargo policy for the journal or find it directly on the journal webpage.

PERSONAL SHARING

A regular question is: should we publish in open access or closed journals (without open access and without publishing fees)? I consider first closed journals where publishing does not require any fees so that I can save grant money. I try to keep my publishing budget for specific cases, which are: (1) publishing in top open access journals, typically the Nature journals, (2) publishing very multidisciplinary work that cannot find in most traditional journals. Typically, projects involving architects, social scientists, and engineers, are very multidisciplinary and cannot match the scope of the journals I am used to. I found that a couple of journals more receptive to this unusal and highly multidisciplinary research are open access, although they may have a lower reputation. For journals with impact factors lower than 5–6, I unfortunately do find that it is easier to publish when the journal is open access.

KEY IMPORTANT POINTS

Publishing strategies:

- Conference papers, literature reviews, postdoctoral research can be written to have some publications when your research group is not yet ready.
- Literature reviews and perspectives will increase your citations; research papers will set you as expert in one field, other types will increase your visibility.
- Make catchy images or videos to share with your work on social media.

6.2 Getting Published

This subchapter covers the publishing aspect of writing, including the process of peer review and how to reach to the editors. Specific tips such as writing the cover letter, discussing with editors at conferences, or responding to invitations to write papers are given.

6.2.1 Submitting to Scientific Journals

After conducting the research and writing the manuscript, it is time to submit it to journals. First thing to do is to make a list of potential journals that could match with the theme and topic of the manuscript. Ideally, it is good practice to write the manuscript with a journal in mind as some journals have specific styles, especially the ones that have high impact factors. If you submit to one journal and it gets rejected, which often happens, then you need to re-tailor the paper to match the format of that other journal. Thus, identifying the journal where you have most chance to get in will save your time and energy.

How to find which journal to submit to and how to build the list of potential journals? You can start with the journals you would like the most your paper to be published in. These journals can be the ones with high impact factor, or the ones with the highest reputation in the field. Then, you can find ideas of suitable journals by looking at the journals that are cited in the reference list of the manuscript. Most journals editors are happy to see that their journal is cited.

Carefully checking the scope of the potential journals is critical to ensure it is suitable for the paper. You can find the scope of each journal on their websites. Sometimes, looking at some of the recent papers published can also give idea of the style and the quality of the papers. Drafting the manuscript in the format of the journal is also necessary as well as to follow other rules, such as about the use of AI or other tools. The journal list you want to build should have 4 or 5 potential journals, which you list in order of decreasing impact factor or decreasing reputation. Sometimes, very reputable journals may not have a high impact factor as it depends on the field. Also, it may sometimes be more favourable to publish timely in a 'medium' journal with the adequate readership, than to publish in a 'big' journal that would take longer time. This is particularly true when you are starting on your tenure-track and need to publish. There are anyways a lot of uncontrollable factors that will make your paper being read and cited or not. Publishing in a 'big' journal does not necessarily translate to a large number of citations.

When choosing the journals to submit, also check carefully if there have any open access fees or if it can be published as subscription with no fee. If you want to publish open access, it may increase the readership and the citation number, but you need to ensure that you have the funding. Some universities have agreements with some journals to lift these fees. If you decide to publish under subscription, check the requirements from your funding agency for this research. Most funding agencies require the results of the research to be made available open access. Therefore, you can also submit some versions of the manuscript on open access repositories such as the one from your university or other independent repositories like ArXiv, ChemRxiv, etc.

PERSONAL SHARING

I find I have more chance to publish in a journal if I have published in it before, especially for 'big' journals. Editors see how many papers you have submitted to their journal and how many they have published. This means that even if you were a collaborator on a good paper, you will be 'in their system' and this will slightly facilitate further publishing with them. Being 'in their system' can also lead to being asked to review for them, which can be showcased in your tenure dossier as being expert in a certain field. Initially, I accepted to review for a wide range of journals. Editors can see if reviewers accept their reviews quickly and deliver the review in time, which I do believe helps in getting invited to review papers in high impact journals. Reviewing papers takes a bit of time but also is a good preparation to learn what reviewers and journals are expecting.

KEY IMPORTANT POINTS

Submitting to scientific journals:

- Make a list of potential journals you are targeting.
- Craft the paper for these journals.
- Think strategically about which journals would be best for your research at that time.

6.2.2 The Cover Letter

In order to submit your manuscript, you have to write a cover letter to explain how the manuscript fits into the journal, how it is new, and why the journal editor should send it to peer-review. If you have not written a cover letter before, you can ask your past supervisor or find the cover letter they submitted in the journal's web portal.

An example of structure for a cover letter can be the following:

- Paragraph 1: State the title of the manuscript and what kind of manuscript it is (article, review, etc.)
- Paragraph 2: Explain what the key novelty of the manuscript is and what it is about, what the impacts of the manuscript are.
- Paragraph 3: Explain in more detail what is unique in your approach or what the scientific novelty is.
- Paragraph 4: State the audience of the manuscript and why it is suitable for the journal you are submitting to.
- Paragraph 5: You can add some potential reviewers in the letter, although most journal platform will require you to input suggestions for reviewers during the submission.

It can be really helpful to meet the editors and get to know them and what kind of papers they are looking for. Some editors are looking for specific content. Talking to editors can be insightful and increase the chances for your manuscript to be sent for peer review. Getting to know the editors can be achieved during conferences, where they may have a booth. You may also look for opportunities to join early career editorial boards or volunteer to peer review. Some journals do this. You may also reach out to senior colleagues or past supervisors, mentioning that you would be interested in having more peer review experience. Being on the other side of the table helps understanding a bit more what the expectations are for cover letters and manuscripts.

PERSONAL SHARING

The cover letter is what the editors read first. They also look at the title, author list, and may open your manuscript to have a quick look at it. Writing a good cover letter is very important. However, your manuscript may still be rejected, even if it was very good. There are many reasons for rejections that may not be related to the manuscript per se and that are out of your control. Nevertheless, if you are provided a reason for the rejection at the cover letter stage, and if that reason is not valid, it is possible to appeal. I have done that before after the editor suggested I submit to a photonics journals, while the work described in the manuscript was absolutely not in the photonics area. My appeal was heard and the manuscript eventually got published in that journal. In another case, I had a manuscript rejected based on the reviewer's feedback, but that feedback was obviously for a completely different paper. It took a lot of emails and generated a bit of frustration to get the editor to realise there has been a mistake. But in the end, it worked out and the paper got published.

KEY IMPORTANT POINTS

The cover letter:

- Present your manuscript to convince the editor to send it to peer review.
- Highlight the novelty of your work and why it is a good match to the journal.
- If you have a good reason, you can always appeal.

6.2.3 The Peer Review

If your manuscript is rejected at the editor level, you can resubmit your manuscript to the next journal in your list, with minor edits to follow the next journal format requirements. If the rejections come with a reason that is not appropriate, you can appeal once to defend your case.

If the manuscript is sent to peer review, there is a good chance of receiving comments for improving and revising the manuscript. If, after the peer review, the manuscript is rejected, do not take it personally and consider the comments and resubmit to another journal. Some journals propose directly forwarding to a more suitable sister journal, which you may consider, or not. I would recommend following your journal list instead, since you took time and effort to make it.

There are new practices for the peer review, which include double anonymized peer review. Whether you choose single or double anonymized peer review is for you to decide. Now, peer review comments are often made available with the manuscript on the journal website. This highlights the importance of conducting the peer reviews respectfully and with scientific integrity.

When submitting the manuscript, you may also propose some reviewers or ask certain reviewers to not be asked for peer review. It can be good to propose reviewers you have met and who you know are interested in the topic. It can also be good to propose reviewers whom you want them to know about your work, and therefore to use it as a tool to build your reputation. Although you may add friends as reviewers, it is also recommended to vary the names as peer reviewing takes time and this may not be considered very appropriate. Whatever names you propose, the editor may choose to send it to them or not.

After receiving the peer review comments, take time to address all the comments and modifications suggested. If one comment is not appropriate or suitable, it is possible to appeal and not reply to it. If there is a good and justifiable reason for not addressing a comment, the editors are usually sensible and will understand. Reading the peer review comments can be tough at times, yet most scientists reviewing papers are genuine and authentic, providing meaningful comments, which indeed improve the manuscript. Reviewers may also give ideas for future research through their questions.

PERSONAL SHARING

I always follow the same format for the 'answer to the reviewer's comments' document. The document starts by stating that I thank the editor, that I have answered all the comments below, and that changes to the main manuscript are in blue (or another colour). Then, I copy-paste the comments and provide and answer for each comment. It is very important to address all the comments, and if possible, to make a modification to the manuscript for each comment. When the reviewer receives your answer documents, they will likely only look at that document, and not read again the manuscript. Hence, it is critical to show clearly that you have addressed everything. Usually, even if the paper is rejected, I ask my researchers and students to address the comments in a revised version of the paper and to resubmit elsewhere. At the end of the day, there are so many journals available that for sure we can find one where it can be published.

KEY IMPORTANT POINTS

The peer review:

- Propose some reviewer's names.
- Address all the comments.

6.2.4 Other Strategies for Getting Published

To increase your chances to publish in top journals or to increase your publishing output, you can consider several other strategies. Some of them are suggested in the following, but you will find your own appraoches as well.

One strategy is responding to invitations to submit a paper. Indeed, you may receive invitations to submit a paper in certain special issues. Although a lot of invitations may come from predatory journals, some come from very high standard journals. Submitting your paper to these issues can increase the chance to pass the editor and be sent to peer review. Some of these issues may also be targeted for early carrier authors, which can be a good opportunity to stand out. You can also look for those issues and propose a paper to the editor without waiting to be invited. It is a good idea to propose to the editor first and not to submit directly in case they have specific expectations.

Another strategy is to propose to the editor a paper idea. Some papers, especially reviews or perspectives, and in some journals, are only by invitations from the editors. If you have an idea of a suitable review paper for a specific journal, you can write a proposal and submit it to the editor to inquire whether that could potentially be a good fit for the journal. They may accept it and even provide suggestions about the content of the review. If the proposal is not accepted, you can send it to another journal. A proposal for a paper would look like a cover letter, except that you have not written the paper yet. Sometimes, it can be good to present the structure of the paper, such as the titles of the sections, and to indicate who will be the co-authors. Therefore, it is a good idea to have a rough draft ready before submitting the proposal.

Other strategies to get to know editors and to publish can be to write outreach articles, opinions or news articles for journals of societies, or even scientific outreach magazines. Indeed, while you want to write, editors want to publish too. Hence, you can again submit brief proposals about articles and see what happens. There are plenty of journals, it is very likely that you will find one in which your paper can fit. You can volunteer to write short news reports or become a member of a society to get to know about these opportunities. The approach is similar if you want to write a book, such as the one you are currently reading. A proposal was first sent out to several publishers before finding a good fit.

PERSONAL SHARING

I got into writing news articles for one journal from a scientific society by volunteering as a student during one conference from that society. This led me to get to know one editor. Then, writing the news articles made me get in touch with the teams who wrote those articles. It not only broadened my knowledge, but also my network. Later, I could propose to write other types of outreach articles such as historical columns. A few years later, I was invited to join the editorial board of that journal, which is a great way to connect with other researchers and potential collaborators.

KEY IMPORTANT POINTS

Other strategies for getting published:

- Respond to invitations to write.
- Propose your own papers to editors.
- Volunteer and write broadly.

6.3 Practical Tips on Writing

Excellent paper writing requires excellent storytelling and illustrating figures. This subchapter provides practical tips on writing research papers for top tier journals, emphasising the importance of crafting intuitive, logical and eye-pleasing figures, and of choosing the appropriate words.

6.3.1 General Rules on Writing

There are many papers and books you can read and workshops you can attend to get some tips for effective scientific writing. Likely, you are already familiar with the process. This subsection provides a more general overview on paper writing, more specifically from the point of view of an assistant professor, which might be slightly different from the point of view of a student or a postdoc.

First, your papers should now be tied to a grant, where you stated you will research on a certain topic and publish a certain number of papers. From this research grant you have made a list of few paper topic ideas that you would like to write. Knowing in advance what you want to write about can make the research and writing process faster.

Secondly, as an assistant professor, you may now engage your students and researchers to do the research and write the paper. Hence, you should engage them at each stage of the writing, from the research idea to the crafting of the story and to the final draft. Discussing your research idea and paper idea with someone else, even if only a student, can help having a critical view and pinpointing the exact novelty that the paper addresses. In addition to the papers written by students and researchers supervised by you, you may also do your own writing, for example for conceptual papers or reviews.

Thirdly, in order to identify the topics and the key novel elements from your research in a broad scope of work, it will be necessary for you to stay on top of the literature. This means that you also need to make time to read, maybe starting with science news or reviews. You can also gain knowledge about the current state of the research when attending scientific workshops and conferences. If your work is within a narrow community, you may follow closely what the other researchers in that community are publishing. If your work can expand across multiple fields and domains, then you should follow key trends in research in these fields.

Reading, thinking, and writing takes a lot of time and energy. You will need to schedule and find the time and circumstances most favourable for doing these, which may mean finding a good location, blocking some time, etc. Keeping records of literature in well-ordered files or using automated software can help. You may also develop your own strategy for remembering the key elements and taking notes effectively.

PERSONAL SHARING

Before writing a paper, or while the research is on-going, I like to download 10 or 20 papers from Google Scholar based on certain keywords regarding the project I plan to write about. Then, I take notes in an old-fashioned way, with pen and paper, in a physical notebook where the date is written. I add a coloured bookmark with the topic of the papers so I can find it later. Then, for each paper I write the name of the PDF file I gave to it, the date of publication, the title, and then some relevant notes. After reading all the papers, I go back through my notes, and highlight the important or useful piece of information. Using paper notebook avoids battery issues or failing to relocate the documents. I find it best to look at papers published in the past 3–4 years to get to know what is going on currently in the field.

KEY IMPORTANT POINTS

General rules on writing:

- Identify what your research papers will be from your grant proposal.
- Engage your students and researchers in the writing.
- Stay on top of the literature.
- Schedule accordingly.

6.3.2 Writing for Top Journals

Scientific papers, especially in 'big' journals, tell a compelling story. They are more than a list of facts and data. The results are contextualized. Those papers usually have clear and appealing visuals, and use words that are straight to the point, which make them provide a large quantity of information in a rather short format.

In a simplified and summarized way, there are 5 key components that make a paper a really good paper:

- The idea: The idea proposed by the paper should be novel and impactful. Usually, the best ideas are the simplest ones too.
- The experimental method: The tools used for producing the results should be creative, novel, maybe unusual or high-tech.
- The results: The results of the research, for example the properties obtained should be higher than that found in the literature, or more diverse, unique, outstanding.
- The theory: There should be some theory available behind the results to explain them and to allow prediction. The theory component is necessary for other people to apply the tools and approach to other context and systems.

- The application/demonstration: There should be a clear demonstration or illustration of how the findings from the paper can find significant meaning or application into the world, beyond the research laboratory.

Generally, a manuscript ticking the 5 components could make its way into journals with impact factor higher than 10, whereas ticking 3 or 4 only, maybe for other lower impact factor journals. As we discussed earlier, publishing in 'smaller' journals may also be very valuable for the research and your tenure package.

PERSONAL SHARING

I found it helpful to inform my students and researchers about how to write research papers by conducting a workshop with them. The '5 elements' rule makes it easier to select the type of journals we can submit our work. Along with the workshop, I also developed my own guidelines on writing papers, where I review at each stage their 'storyline' before they start writing the entire paper. It saves time and frustration if we (all the co-authors) agree on the storyline before starting to write the entire story. Sometimes, I craft the storylines for my students, especially the first papers of PhD students so that they can see the process. In some cases, when the researchers have left or just got interesting data that we want to publish quickly, I also do the writing of the first draft for them, and they can fill up the technical gaps. I think research is teamwork and being a student does not mean that the professor does nothing.

KEY IMPORTANT POINTS

Writing for top journals:

- The ideal paper has 5 key components: idea, experimental method, results, theory, application.
- A scientific paper tells a story.

6.3.3 Data Visualization

In addition to the content (the results, the discussion), the form of the paper is also important, in particular the figures. In high impact journals, the number of figures is usually small, of 4 to 5 figues only. The figures are figure panels that group several images and graphs supporting the same message.

How to make visuals that are appealing and telling a story by themselves can be critical for several reasons.

First, figures are particularly important because they will show up in Google image when someone keys in certain keywords related to the research. Having appropriate figures can help your paper being clicked on and therefore being read. Clear and appealing figures can be reused more easily in media, increasing your impact.

Secondly, having good figures can be critical for the review process. After reading your cover letter, editors are likely to have a peek at your figures before deciding to send the manuscript for peer review or not. The figures show the data that support the findings of your research. Reviewers may just browse through the figures before deciding how good the paper is. Readers will do the same.

How to make good figure panels? One figure panel should have one meaning and be linked to one section of your paper. The colours and the ways the data are represented may influence how the reader understands the data. It is therefore important to guide them with appropriate data visualization. You may support yourself with some software to facilitate the creation of schematics and graphs. You may also ask other students or researchers their understanding of graphs to ensure the data is appropriately visualized. Looking at papers that have figures you like can be a way to get inspiration for crafting your own figures.

PERSONAL SHARING

You can craft the entire story using figures first, without any text. I start drafting the storyline for papers by placing all graphs and images in one power point slide deck. Then, with the other co-authors, we start organizing the images and graphs together depending on their meaning, to finally obtain figure panels. Once we have the figure panels that tell the story. It is very easy to write the paper because it is mostly describing in more details what is in the figures. It also ensures that there is strong data to support any statement in the paper.

KEY IMPORTANT POINTS

Data visualization:

- Crafting clear and easy to understand figure panel is very important.
- Just looking at the figures of a paper you should be able to understand the story.

6.3.4 Words

Words are very important. While science means the data should be the most important, for publishing papers, attention should be paid to the words that are used. Some words have different meanings for people in different fields. Examples of such words can be 'texture', 'orientation', 'model', etc. It is therefore critical to employ unambiguous words or to give their definition. Some words are not recommended to use, such as those overemphasizing your work: 'novel', 'innovative', 'unique': there are your own subjective views. Other words are typical of artificial intelligence (AI) use like 'delve', or 'underscore', which should be avoided. (The use of AI to write papers should be avoided altogether). You should also use the words adapted to your readership, which depends on the journal you intend to submit to.

Careful choice of words will also help you write precisely, concisely, and convincingly. Specific examples are given in the following:

- *Write precisely but concisely.*
  Can be commonly seen (4 w) → is common (2 w)
  Greatly impact (2 w) → increase (1 w)
- *Remove unnecessary words that do not carry necessary meaning.*
  In order to (3 w) → to (1 w)
  The aim is to realize the manufacturing of X (9 w) → The aim is to manufacture X (6 w)
  This paper provides a review of (6 w) → This paper reviews (3 w)
- *Use the verbs in the active form.*
  PVB was added and mixed using milling balls (8 w) → PVB was added and ball milled (6 w).
  Classified between two groups (4 w) → divided (1 w)
- *Cut long sentences in 2 or merge 2 long sentences into 1*
  The degree of the effect of magnetic fields depends on various factors and one of them is the particle size (20 w) → The degree of the effect of magnetic fields depends on various factors. One of them is the particle size (19 w)
  As epoxy is used, the value of x is 2. So, using equation 1, … (14 w) → The value of x is 2 for epoxy, resulting to … (10 w)
- *Remove repetitions.*
  SEM microscope (2 w) → SEM (1 w)
- *Be honest but positive.*
  Simple modifications of the system → modifications of the system.
  To make → to create
- *Place the most important point as the subject of the sentence.*
  An alternative approach is to formulate printable materials → Printable materials are a promising approach.

- *Use active verbs*
  PVP was used as the binder (6 w) → PVP bonded (2 w)
  It is expected that the average value (7 w) → The average value is expected (5 w)

PERSONAL SHARING

I find that some students do not know how to write and tend to write as they talk. Being a non-native English speaker may be one reason for the challenge in writing. Providing specific guidelines such as those listed above can be very helpful for them to develop efficient and appropriate writing skills. When providing feedback on their first draft, it is very important to be able to tell them what needs to be improved and how. This will save time for you later as they pick up the skills. Asking your students to peer review papers with you is also a great way to train them. If I am asked to peer review a paper that is relevant to their research, I sometimes ask them to do the review with me. Some journals allow you to do this and to declare it in the peer review, which is nice. But for others, I still do it while keeping the confidentiality.

KEY IMPORTANT POINTS

Words:

- Choosing the right words is key to be understood, to be concise, precise, and clear.

6.4 Conclusions

In this chapter, we have reviewed strategies to publish and to increase the impact of your publications. We have also covered practical aspects of writing cover letters and drafting papers. Publishing, again, is still one key assessment criterion for assistant professors in any field. It is important to learn how to navigate the publishing landscape confidently. Because there are many journals and many types of articles, there are plenty of opportunities to try and see what happens. The skills for writing impactful papers can also be used for any other types of writing, from grant applications, to technology disclosures or your statements for your tenure dossier.

Additional References

J.B.P. Soares, Publication tips: how to write scientific articles that master the publication process and communicate your ideas, Can J Chem Eng 100 (2022), 2718–2731.

N.M. Glover, I. Antoniadi, G.M. George, L. Gotzenberger, R. Gutzat, K. Koorem, P. Liancourt, K. Rutowicz, K. Saharan, W. You, P. Mayer, A pragmatic approach to getting published: 35 tips for early career researchers, Front Plant Sci 7 (2016), 610.

J. Beall, Predatory publishers are corrupting open access, Nature 489 (2012), 179.

R.E. Schaak, Increase your impact: writing tips to reach a broader audience, ACS Nanosci Au 2 (2022), 250–252.

P.E. Bourne, Ten simple rules for getting published, PLoS Comput Biol 1 (2005), e57.

7

Teaching

7.1 Teaching at the University

Universities are institutions of higher education. Universities are different in their structures, goals, and student populations to smaller pre-university educating entities. Professors at universities are engaged in teaching. Some of the teaching activities are part of a specific curriculum, but some are not.

7.1.1 Specificities of University's Teaching

Teaching and learning at the university are different in nature to teaching and learning in high school. Entering university for their undergraduate degree often coincides with the beginning of adulthood for the students. They expect to not only learn new knowledge, but also to enjoy social and extracurricular activities, which should lead them to finding their path. As such, university teaching generally makes the students the owners of their own learning, with many opportunities for them to choose which courses to follow. Although we all have different reasons for entering an undergraduate program, a large majority of the students still have the primary goal of getting their degree first, and quite a few may be there because their parents asked them to. Only a few will continue with postgraduate studies and further. Despite this disparity, the content delivered to them should be mindful of these motivations and address all students.

Furthermore, universities are also larger entities in size than high schools, with cohorts of students that may reach 300, 500 or more students per year and per department. The students joining the program may likely have studied different topics prior to the university, coming from different kinds of schools. The student population may also comprise local and international students. Some students may also be enrolled full-time and there may be part-time students that pursue their studies in addition to having a full-time job during the day. Universities are divided into schools and departments, and some courses may be given across departments leading to groups of students following different majors sitting in the same room, which further adds diversity to this melting pot of students.

DOI: 10.1201/9781003635116-7

Teachers at the university are lecturers and professors, including assistant professors, who spend a lot of their time doing other things than teaching. The motivations of the university instructors therefore can also be broad, from graduating students to encouraging them to pursue an academic career or exploring another career path. Some other lecturers may be hired to do only teaching, while some may do part-time teaching and have another job. Overall, teachers at the university are here to 'help' the students do well and realize their goals.

PERSONAL SHARING

I never studied in a classroom having more than 100 students at once. The first time I was in front of 300 or more students felt very scary. Especially because they do not sit quietly waiting for you to talk. Luckily, the only time all the students come is when there is a quiz or an examination. Hence, as a lecturer, we have a lot of authority in these situations since we are the ones doing the grading. Using the microphones and displaying the slides are signals for the students to quiet down. University students overall are quite well-behaved, so even in large groups, although it is intimidating, it is manageable. However, it took a while for me to relate to what it is to be a student in a class with 300 other students.

KEY IMPORTANT POINTS

Specificities of university's teaching:

- The student population is very diverse.
- There are a lot of students.
- We are all here for different reasons.

7.1.2 Teaching Content Part of the Curriculum

University's curriculum is well defined and follows certain standards. Every 5 or 10 years, the university's departments go through the process of accreditation which is a formal evaluation and review of the quality of a university program. Because the program has to meet certain standards, if you are asked to teach an existing course or asked to develop a new course, you will need to define the specific learning objectives in addition to the detailed course content. Also, ideally, the course developed should fit and be aligned with the entire curriculum.

There may be various modes of teaching: lectures, tutorials, laboratory experiments, projects, etc. The styles of teaching can also vary a lot, from in-person teaching in lecture halls, to online teaching, recorded videos, group teaching, etc. One popular current way of teaching is called flipped classroom, where students study the course material at home, at their own pace, and come to the class to discuss what they have learnt at home. This enables more personalized learning, with more time to go deeper into the concepts as well as more efficient learning in class.

At the university, classes may be taught by one single lecturer or shared with several lecturers and teaching assistants. This may demand a little bit of coordination. There should also be an online platform to share course material. The type of assignment should also be set clearly and can follow different modes, such as continuous assignments, final exams, home assignments, presentations, etc.

PERSONAL SHARING

Here is an example of a course proposal that did not fit at all the overall curriculum and the general teaching style at the university. I proposed once to develop a course on colloids (a type of microparticles suspended in a liquid) which I wanted to be for small groups of students and project-based, with each group of students studying one particular phenomenon related to colloids and developing the experiments to explore it. Then, the students prepare a report and a presentation to share about it with the entire class. This was aimed for a class about 50 students, divided into groups of 2 or 3. I was told that the course would be too complicated for the students, too challenging, and too much work for me. The feedback was probably accurate in that specific context.

KEY IMPORTANT POINTS

Teaching content part of the curriculum:

- Develop quality according to the accreditation requirements.
- Choose your teaching mode and style.
- Make sure to align with the general trends at the university.

7.1.3 Teaching Content Outside of the Curriculum

Apart from the core curriculum, it is also possible to engage in various other types of teaching which may be more flexible and allow more creativity, especially if it is not linked directly to student's exams and degrees. For example, as part of a teaching or research grant, you may develop a workshop to teach about a certain concept. The workshop can be used to gather data for research, or simply be a means of doing outreach and sharing about a new topic that may be related to your research.

These kinds of sporadic teaching events may be more closely related to your specific areas of interest and can involve your own researchers and students as well. They may be targeted for students at the university, but also could be for students outside the university and at various levels. Because there are not tight to a specific curriculum, there is a large flexibility. These teaching activities are playgrounds to explore and improve teaching skills.

PERSONAL SHARING

I have been doing quite a lot of these 'outside of the curriculum' kinds of teaching. I developed workshops as outreach for a research grant where my researchers and me taught pre-university students as well as members of the general public about the fabrication of materials using fungi. This was a means for the researchers to deliver lectures which is a good addition to their CV, as well as to bring general awareness to the public. The workshop took half a day and comprised lectures as well as hands-on experiments and brainstorming. Another workshop I developed was part of a teaching grant with seven other professors and lecturers at the same university but in other departments. We developed a 2-day course about using artificial intelligence for writing literature reviews. The course included lectures, writing assignments, and discussions. The goals of this workshop were to collect data and the proper ethical protocol was followed to recruit the students and analyze the data. Finally, a last workshop I took part of was not related to any grant but to my personal interest in biomimicry where I collaborated with an overseas professor to develop a course. Because there is no specific expectation from this collaboration, it enabled us to really explore and improve over the years. For example, we have been including more cultural elements into the workshops, field trips, students from different backgrounds, experiments, etc.

KEY IMPORTANT POINTS

Teaching content outside of the curriculum:

- Allows more flexibility and exploration.

7.2 Preparing and Delivering Courses

Before delivering the course content to the students, the course materials and the assessments should be carefully crafted. This section discusses these three points.

7.2.1 Preparing the Course Materials

Whether it is a course material for an existing course or a new course, it is advised to craft your own version of the course and not to directly take existing slides. Crafting your own materials will ensure that you master them. However, it is likely more efficient to take inspiration from existing course material, either from another faculty teaching the course before you, or from external resources online.

It can be very helpful to build the course material on an existing textbook. Not only it will ensure that everything relevant is covered, which can be good if the topic of the lecture is not something you are an expert on, but it will also be an additional reference for the students to consult if they need more information or details.

A typical course material should be very clear and organized. The course overview with the intended learning objectives of the course, the introduction of the topic and its importance, who is or are the lecturers, and the course schedule should be presented first. The modes of assessments and how the students will be graded should be introduced as well. Having a detailed work plan and target dates for the assessments will help the students prepare accordingly. Across the lecture, it is good to remind the students how what they are learning is related to their assessment as well as which parts are important, what they should remember and why.

Besides the content, it is recommended to explore various formats to engage the students, for example having slides with text, but also images, videos, etc. The attention span of students is the highest in the first 10 minutes. If the class size permits, planning activities for the students such as discussions, readings, comparisons, etc. can also be helpful. Even if there are activities and less formal modes of teaching, it is very important to have it carefully planned beforehand, especially the time allocated for each activity, as the time schedule should be followed. Furthermore, back-up strategies should be also thought of in advance, for example in case of a lack of student's participation or responses. Depending on the culture of the university, online participation using QR codes and platforms like Wooclap, Mentimeter, etc. can engage the students without leading to uncontrollable discussions. Having these activities may require a little bit of practice.

PERSONAL SHARING

Here is an example of teaching styles I tried which worked in one context and did not work in another context. During the COVID times, when we had to teach only online, I used Wooclap to run some little quizzes during my tutorials on fundamentals of materials engineering and manufacturing. The quizzes were anonymous multiple-choice questions set using Wooclap. I asked to get at least 70% of the students to send me a response, even if it was not correct. I would wait to get these 70% answers. Then I would explain the correct answers. I was doing this in the tutorials, breaking down the exercises into a lot of these small quizzes. The feedback I received was that it was fun, engaging, and helpful as I would also explain more when a lot of the answers received were incorrect. Because of the anonymous nature of the quiz, students participated heavily. However, I tried once the same approach in person in a classroom after COVID and it did not work at all, and I am now back with slides and me talking through the exercise solutions.

KEY IMPORTANT POINTS

Preparing the course materials:

- Having a reference textbook can be very helpful.
- Develop the content but also the form.
- Try it out and improve based on the feedback received.

7.2.2 Preparing the Assessments

To be able to prepare the students for their exams, the assessments for the course should be developed at the same time as the course is developed and delivered. Only by knowing what we want the students to know can we emphasize to them which are the important key concepts or pieces of knowledge that we want them to know.

Forms of assessment vary and how you will grade them will also vary. There could be multiple choice questions, short answers, assignments at home, reports, presentations, closed-book or open-book examinations. For a start, it is advisable to follow the common practices at the university when setting up assessments. Secondly, the modes of assessment should be adjusted to the typical class size and communicated clearly to the students early on. Likely, they would already know about the assessment when enrolling in a course.

After deciding on the form of assessment, you should craft the assessment to be easy to grade for you, especially if it is a large class. How to make exam papers easy to grade? For example, for continuous assessment and small quizzes throughout the semester, multiple choice questions can be the way to go. The results for each question may also be gathered in a summary table that the student has to fill in, which avoids having to open and flip papers during the correction. For long exam questions, you may design the question so that you are looking for a particular answer. For example, by asking them to draw a curve or make schematics, or to do a calculation. If the answer is in text form, you may look for specific keywords. However, sometimes you may need to adjust the severity of your grading depending on how good the overall answers are.

Because examinations at the university are a very important and controlled you should know that your examination papers will be checked by other professors to ensure they are free of errors and clear to the reader. It is very important to have clear and neat exam questions as making a correction in front of 300 students can be embarrassing and lead confusion among the students. It is also very important to keep the examination papers confidential and to change the examination questions every year.

PERSONAL SHARING

I like to have various types of questions in my examination final papers. Usually, I have some basic questions close to the lecture, where the students have to make some drawings or schematics. Other questions are more discussion where there is no right or wrong answer, but depend on the arguments they bring forward. These questions allow me to find points for students who did not do so well in other questions. Then I also have some calculations. Calculations are generally hard to grade as some final results may be incorrect but the way to go through them was correct. Depending on the difficulty of the calculations, I sometimes give points for good intermediate level. I find that in general, grading final exam papers consists of finding ways to give students points.

KEY IMPORTANT POINTS

Preparing the assessments:

- Make sure they are free of error.
- Choose your assessment modes and what will be assessed while preparing the course material.
- Prepare the assessment questions to be easy to grade.

7.2.3 Delivering the Course

Delivering a new course for the first time can be stressful, especially in front of a large class. Being very familiar with the course materials and having rehearsed is necessary. At the assistant professor level, very good teaching is not necessary but bad teaching will definitively not be acceptable. Because there is time to progress during the tenure-track, it is important to make experiments in the first year and learn from these teaching experiences to improve continuously. It is good to find your own style. You cannot afford to be very stressed and not confident every time, that would be too tiring. Find your comfortable style. From the student's perspective, they may have expectations of what a teacher should be and it is good to match this expectation with your own twist.

If teaching does not come naturally, you can therefore prepare a bit more at the start and also attend others' lectures to get indications of how to teach effectively, how to manage the class, as well as to be more familiar with the students' culture and behaviour. Because teaching commitment for professors is regular, you need to find your own ways to enjoy your teaching so that you can communicate your enthusiasm to the students.

Besides mastering the content of the course and delivering it confidently, keeping the time is also very important. Students will not allow overtime as they have other classes to attend, and you will need to cover the entire course. Being too fast is also not a good sign and may be perceived as not caring and bad time management, but could be leveraged by slowing down at the end or providing an extensive recap of the course. If the course is not a lecture but a tutorial, scheduling time for questions can also be good practice.

Finally, when delivering the course, it is good to know who is in the room and to remind oneself of the diversity of the students. If this is an interdisciplinary lecture with students from different schools, adapting the content to the background of these students is good. Also, students may have visible or invisible disabilities. Ensuring text is big enough to be read, that the microphone or your voice is loud enough, and that the course materials are available later is also very good to not leave anyone behind.

PERSONAL SHARING

Because universities give a lot of freedom to the students to choose and attend the courses that they want, it is always unknown how many students will turn up in person for lectures. In addition, our lectures are all recorded live so that students can watch them later. Although I always feel disappointed when only 1/4th of the students turns up for my lectures, it is actually better as the most interested students are there. Also, it means they are quiet or may be asking questions. This makes the teaching a pleasant experience. Now that my course is rolling and that I know the content, I find it more and more enjoyable to teach as it is a nice break in the day. Over the years, I modified slightly the content to teach about areas that I particularly like. For example, I am teaching a lecture on polymer processing (extrusion, injection moulding) which I do not find particularly exciting. Those parts I prepared recordings for the students to watch and I reuse the same recordings over the years. But I quite enjoy teaching about the viscosity of polymer melts so this part I keep in person and edit a bit every year. Now I added a small experiment dissolving LEGO plastics in nail polish which I find funny and I think surprises the students (they are probably thinking I am crazy).

KEY IMPORTANT POINTS

Delivering the course:

- Master your topic.
- Make sure you are enthusiastic when delivering the course.
- Keep to the time carefully.

7.3 Receiving Student Feedback

Teaching performance is partly based on students' feedback. What it is, how to prepare for it, and how to receive it are discussed in this section.

Teaching is an activity that is important for the overall performance of an assistant professor. One main way teaching is evaluated is through students' feedback. In a similar way that you need to prepare the students assessment when preparing the course materials, the students' feedback should also be prepared ahead of delivering the course. You should know how the students will provide feedback to you and on which criteria.

Informal feedback could be also thought directly during the lectures. First, by looking at the students and sensing their level of engagement. For example by asking questions, or asking them to do a task in class. Again, anonymous methods such as Wooclap or Mentimeter can be used to assess the level of engagement and understanding of the students and allow to rectify live what was not conveyed successfully. Another way to get direct and informal feedback is also to ask them directly.

Most importantly, the official students' feedback received should be taken constructively as a means to improve future teaching. Overall, students want clarity and responses if they have questions. If you receive negative feedback, it is unlikely that it will come as a surprise to you. Being aware of the students' attitudes during the class is usually a fairly good indication of how they perceived the teaching.

PERSONAL SHARING

In my university, the professor sets the questionnaire that will be sent to the students and serve as the students' feedback. I need to know exactly which questions I set so that I make sure I do that in the class. In addition, proper timing for sending the questionnaire is critical. Once I sent the feedback questionnaire after doing a graded quiz in the lecture. In that quiz, I asked some questions that were covered in the tutorials, but not in directly in the main lecture. I received numerous complaints in the students' feedback about this. Nevertheless, that was a good learning for me: to send the questionnaire before the graded quiz. The other unpredicted side-effect of me asking tutorial questions in the quizz is that now, every semester, I have students asking me which tutorials I will be teaching to join my group as we are several instructors for the tutorials.

KEY IMPORTANT POINTS

Receiving student feedback:

- Student feedback is important for your performance as assistant professor.
- Know in advance on what criteria the students will give you feedback.
- Informal feedback can be received during the classes.

7.4 Managing Challenges

Teaching may come with multiple challenges, which spans from understanding the students' perspectives to solving errors, finding last minute replacements, and coordinating with colleagues.

Challenges in teaching may arise during preparing and delivering the course, setting the assessments, receiving the feedback, etc. We have just discussed previously how attending others' lectures or even watching online inspiring lectures can help with the teaching. We have also emphasized how important it is to enjoy teaching and to be enthusiastic, so finding ways to have fun is also important. But aside from the teaching per se, other challenges may come.

For example, coordination between multiple teachers for one course may be needed to avoid communicating contradictory information. There may also be disagreements between the lecturers on the content of the course, for example if a theory or an equation is controversial. All these challenges resolve by professional and timely communications. In general, it is a rule to never to criticize a colleague in front of the students and to keep any disagreement behind closed door.

Other challenges that can occur are making mistakes either in the lecture content or in the assessment of the students. The way to handle such errors professionally is to communicate clearly and honestly with the students, maybe explaining briefly why the error occurred and proposing a way to correct it. Students are generally understanding. However, it is also important to take time to correct it correctly, in the sense that providing multiple corrections would not appear very professional. Overall, it is good to always double check information before sending to the students to avoid confusion.

Finally, what to do if you are suddenly sick or have to miss a lecture for any reason? If it is a last-minute thing, it may be difficult to find a replacement. Ideally, it is important to inform the students and the other colleagues in charge of teaching in your department as soon as possible. If a replacement cannot be found, there is always the option to redo the lecture at another time, if the students agree, or to provide a recording. Similarly, if the absence is known in advance, replacement or online session or recording can be arranged. Usually, students are agreeable with such arrangements. Students should be informed as soon as possible and, even better, asked for what they prefer (someone replacing or a recording).

PERSONAL SHARING

The biggest challenge I had initially was to adjust to the student's prior knowledge and interest as I was not familiar with the local education system. To address this challenge, I did a little bit of reading on the topic and also joining judging events with pre-university students. This enabled me to better understand how typically students behave, what they have been trained to do before, what they learnt, and the multiple paths that can lead them into my class at the university.

KEY IMPORTANT POINTS

Managing challenges:

- Manage mistakes professionally.
- Coordinate clearly with colleagues and students.

7.5 Conclusions

Teaching is an expected duty of assistant professors. Teaching at the university is different to other types of teaching due to its broad and diverse student populations, which in turn allows more diverse forms of teaching and more opportunities to develop one's own style of teaching as well as scopes of interest. Preparing courses for teaching comes every semester and should be prepared thoughtfully in advance. Indeed, despite the flexibility and diversity, the calendar is there and students need to graduate. Teaching is integrated within the service and research activities and sometimes may confront or instead overlap. Organization and efficient communication are key to make everything work smoothly.

Additional References

https://facultyworkshop.com/howtoplanyourfirstcourse/

G.J. Siedel, Seven Strategies for Successfully Preparing for Your Next Class (August 26, 2022). Available at SSRN: https://ssrn.com/abstract=4389958 or https://doi.org/10.2139/ssrn.4389958

J.E. Lawrence, Teaching large classes in higher education: challenges and strategies, Educ Rev USA 6(6) (2022), 251–262.

C. Mulryan-Kyne, Teaching large classes at college and university level: challenges and opportunities, Teach Higher Educ 15 (2010), 175–185.

S. Cornes, D. Torre, T.B. Fulton, S. Oza, A. Teherani, H.C. Chen, When students' words hurt: 12 tips for helping faculty receive and respond constructively to student evaluations of teaching, Med Educ Online 28 (2023), 2154768.

J. Eison, Confidence in the classroom: ten maxims for new teachers, Coll Teach 38 (1990), 21–25.

8

Work-Life Balance

8.1 Is There a Work-Life Barrier?

Work-life balance suggests that there is clear distinction in space and time between work and other life-related activities. Is it always the case in reality? This subchapter discusses this by giving examples while suggesting some advantages of this aspect for creativity and inspiration, and pointing out the associated challenges too.

8.1.1 What You Can Expect

'Academic life' is a very vague concept, which differs from one person to the other, and is influenced heavily by the local and personal context. But your expectations may be different from reality. Academic positions like assistant professor on a tenure track are highly competitive, and there may be some lack of clarity about the expectations to attain promotion. As a PhD student or postdoctoral researcher, there was similarly hardly a strict delimitation between what is studying and what is 'work', and bringing home homework was considered expected and normal. As a new assistant professor, you may be ambitious, excited, and have great expectations, which is great, but 'sky is the limit' may not be sustainable, and you need to pause and consider what is good for your wellbeing. Below are some details about what you can expect.

When signing your working contract as assistant professor, there may be specific working hours where you are expected to be in office, as for any other employee in the university. Very likely you will find yourself not following these hours, and your colleagues neither. In addition to this, remote working may also be a possibility that may further blur the barrier between work and life. Remote working may be due to conference trips or other personal reasons. Beyond the flexible working hours, the aim to attain tenure will add to your inner passion and motivation, leading to a feeling that you should spend (and maybe want to spend) 100% of your working time on research, and the rest, teaching and service, comes on top of that. It is also well known that research is a long-term thinking process, where ideas and questions may run in the back of your mind, even when you are doing something else. Don't we get our best ideas when taking a stroll in nature or under the shower?

DOI: 10.1201/9781003635116-8

Depending on your university culture, you may also find out that colleagues, students, and including undergraduate students, write to you in the evening, or on the weekend. Even managers and other administrative supportive staff may extend their working hours over the weekend and at night. Colleagues on travel overseas may also email at strange times due to the time difference. You may have collaborators overseas, in a different time zone, leading to meetings falling either very early or very late. If you have meetings with people from several time zones, very likely it will not happen at a decent time within the 9 am to 5 pm window. In Asia, it will most likely be after 9 pm.

You also may be requested by your department to attend outreach events to promote your university and school. Such outreach events generally occur on weekends and during holidays. Students and student clubs also have extracurricular activities and may ask you to participate to give a talk or join a dinner. This will likely happen on an evening or a weekend. Even when attending conferences, although they are technically business trips, they likely start on Sunday, with all sorts of activities morning, day and night, during the entire conference week. All these extra hours will be taken from your own good time. There might also be part-time teaching. Part-time students have their job during the day, and then attend their classes in the evening. Someone has to teach them in the evening…

Finally, you will very likely engage in other activities such as the peer-review of papers or grants. You will also write grant proposals and research papers. You will read papers and keep yourself up to date with the literature. You will grade exams with a strict deadline set by your school. You will prepare last minute reports for defending your grant. You will review and grade student's reports and prepare lectures. Fitting all these activities in a day is possibly feasible. At least now you know about it in advance, the variety of activities that will be required from you to get organized accordingly.

PERSONAL SHARING

I was not prepared to receive emails from undergraduate students late at night as they are stressed by their exams. I was told that answering students right away is good as they will give good feedback on teaching which is used for evaluating the teaching performance. But avoiding answering students right away, especially when it is during the night or over the weekend, I find that issues usually resolve by themselves. I send them a quick note on the next working day to show that I still care, and they probably get the hint that they did not write at a good time. Undergraduate students (or other types of students) also sometimes just walk into your office without arranging a meeting before. I do not like this, and I have to send them away, explain the process, and re-arrange a meeting properly.

KEY IMPORTANT POINTS

What to expect for work-life balance:

- Academia context blurs distinctions between work and life.
- University culture may further contribute to this.
- Be aware of what might come to you and what you can deliver.

8.1.2 Strategies for Work-Life Balance

Work-life balance is an important aspect of your career to consider. Work-life balance means being able to comfortably allocate time for your work activities and for your other activities. There are many relationships between job satisfaction and everyday life satisfaction which have been reported in the literature. It is therefore key to take all aspects of both job and personal space into consideration. It is particularly important as academic positions are aimed to be long-term, hence more like running a marathon than a sprint. Strategies to create and keep a suitable work-life balance include to learn how to say no, being choosey, and to embrace flexibility.

First, learning to say no to additional tasks, or requesting deadline extensions may not be so easy, especially as a starting assistant professor. Saying no, however, comes more easily when you have a good reason for it. For example, you have another appointment at that time, or you have already booked holidays, or agreed to another important commitment. Although you do not need to state the exact reason why you are refusing something, you can mention your other commitments. Saying no without feeling uncomfortable therefore implies to carefully choose what to say yes to. You may decide in advance the number and the nature of the activities you want to commit to. Once your list is full, you can refuse all other requests. Therefore, being choosy is important.

You can select what is really meaningful to you. For example, you can accept an extra activity if you respond yes to these three questions:

- Can I add this to my CV? Will it benefit my tenure-track dossier?
- Is it fun? Do I want to do it?
- Is it worth the time and effort? Do I actually have the time and energy to do it?

The challenge lies in identifying opportunities which are useful in the present or in the future. Maybe one way to avoid the Fear-Of Missing Out (FOMO) effect is to realize that there will always be opportunities. You can allow yourself some time to take a decision before replying to any request.

To help with your scheduling and to create a balance between work and life, you can also embrace the flexibility of working time that academia gives you. You can carefully schedule and book your 'life' activities in a fixed or a non-fixed routine. If these are set in your calendar in advance, then it becomes easier to say no to extra activities that might be concomitant to these ones. You know it is very easy to forget, hence the calendar can serve as a reminder for yourself. What to include in your calendar can be your personal activities, but also what fits your character and way of working. For example, if you prefer to work in the morning or in the afternoon, you can arrange your calendar to match your own personal way of working to be more efficient.

PERSONAL SHARING

I attended a couple of outreach events with high school students and parents in my first year. I honestly found they had 0 value, 0 impact, and so much time and energy. Probably being a foreigner who does not even know the names of the schools was a reason for the lack of connection and interest at these events. Now, I have (several times) intentionally taken a leave day to have the excuse that 'Sorry, I am on leave that day' to avoid participating to some outreach activities during the week that I could not find any other excuse to decline. I also often say 'sorry, I have other commitments' when my commitment is going to a gym class in the afternoon or just to the hairdresser. People do not need to know the details, but academia is flexible enough that even taking an extra leave and working that day means I have no shame in taking one day off unofficially. I also usually refuse any extra activities on the weekends or in the evening, unless it really sounds interesting or fun. For example, we had women star scientists coming to give talks on a Saturday afternoon. It was a great event with high quality content and friends and colleagues there. I went another time to support another women-related outreach event which was fun because I knew the organizing team, etc.

KEY IMPORTANT POINTS

Strategies for work-life balance:

- Say no but do not give too many details.
- Add your personal commitments to your calendar.
- Select what you say yes to.
- Enjoy academic flexibility.

8.2 Recharging Needed

Taking breaks is key for long-term success. This subchapter discusses the key role of holidays and how to take small or long breaks in creative ways.

The main challenge in academia and as a professor on tenure-track is that although there are certain goals you need to attain, you are mostly setting the bar and the aims to attain, yourself. While it is critical for your success to find out what the exact expectations for getting tenure are, you also have your own agenda. Given how competitive academia is and how blur is the target, it may be challenging to assess how you are performing, not only to your own standards, but also by the standards of your university. Likely, you will have yearly performance discussions to understand where you are, but you should also be proactive in taking breaks in order to avoid burn out.

You can find creative strategies to take short daily breaks, as well as long breaks. Usually, any break will not only help to recharge, but also help to refresh and get new ideas, inspiration and motivation. Short daily breaks can be easy to create such as going for lunch, lectures, walks, and doing other personal activities. Short breaks can be pure refreshing of your mind and body, or be productive, such as casual meetings with colleagues. While relaxing, informal chats with colleagues can create opportunities for collaborations, or give you insights about how the university or department is functioning. Therefore, such friendly interactions can be a good opportunity to get some tips.

If your working day involves sitting at your desktop for long hours, small regular walks can avoid physical tiredness and other physical discomfort. Standing up and working for 2 minutes every hour, as well as relaxing your eyes by looking at something else, are reported to have instantaneous benefits and are easy to implement. Sometimes, short breaks can also be day-dreaming or short nap in your office... It is a good idea to arrange your office to be a pleasant place for you to feel comfortable. Some people buy a sofa, others a standing desk, or simply a fan. You can include plants, pictures, funny objects, etc. Small investments to make your office more comfortable and more suitable for your productivity are definitely worth it. You can also ensure that you choose your office location or to change it if there is an issue, such as noise or temperature problems.

You will also need, and want, to take long breaks. They can be productive breaks like conferences, or workshops, although these may also be very tiring depending on your level of commitment, social effort, and how tiring is the travel to get there. You therefore want to choose the conferences and workshop you attend carefully. Conferences can be breaks into your routine and the change of landscape likely will refresh your mind. However, traveling and meeting new people may be relaxing or stressing, and you may want to take extra care to self-evaluate your own tiredness and stress levels.

Conference trips can also be combined with a few days off to take advantage of the journey, or to recover from the journey if it is very far.

Long and real holidays are also breaks you will obviously want to take to unwind and to connect fully with your other 'life' activities. As a faculty, you probably will have a decent number of paid holidays to take. It is strongly recommended to take all of the days or nearly all. If you manage to go on holidays without your laptop and can disconnect fully, it is even better. You will come back with more energy, ideas, and motivation. Usually, it is said that you need at least a long 10 days break to most effectively disconnect (Figure 8.1).

FIGURE 8.1
Short and long breaks are much needed.

PERSONAL SHARING

To avoid the double jetlag during conferences and when coming back, I often try to keep my normal 'home' hours. This has the benefits of delivering a fresh lecture, enjoying the hotel, but makes me quite anti-social, especially when the time difference is 12 hours… People are generally quite understanding. I try to schedule my social interactions in advance, such as asking people to chat and setting at time, to still enjoy that part. I also never take luggage to avoid the stress of losing my luggage and plan sufficient time for flight connections to enjoy the lounge before long flights (and avoid rushing and stressing if one plane is delayed). Developing habits for travel can save a lot of stress and if you are a frequent flyer, having the membership of an airline can come with a large a number of benefits, from easy check-in to upgrades to larger seats, lounge access, free internet connection, etc.

KEY IMPORTANT POINTS

Recharging needed:

- Take short breaks.
- Take long breaks.

8.3 Having Fun at Work

Given the time and commitment of a professor to their job and institution, joy, fun, and enthusiasm are keys for sustaining passion and motivation. This sub-chapter employs itself at highlighting ways to find joy and fun in the everyday work activities.

Having fun at work deserves its own section. If your job is not fun in some ways, then it may be time to reconsider it. The job should be fun and make you happy, despite the challenges and hurdles. You should identify which parts you enjoy most, and which ones you do not. Is it writing? Is it teaching? Is it building teams? Is it setting up new laboratory equipment? Is it organizing workshops? Etc. Then you should reorganize and reconsider your activities to have most of the ones you like. Feeling happy and enthusiastic will also have a positive effect on your team and inspire and encourage your researchers and students. Enthusiasm is contagious.

To make the most of what you enjoy and what is useful for your aims, and getting the tenure, you can choose your projects, teaching, and service according to what you really like. As mentioned in the previous section, asking if the activity is fun should be a decisive factor for saying 'yes' or 'no'. Attending a conference because it is in a country you have never been can be fun, on top of being interesting and useful for networking. Accepting to being a judge at a science fair can be fun because it is with high school students, in a museum, with different people, in addition of being useful for you to better understand who are the students that join the university and what they know, as well as being a 'short break' from your other research activities. Going to a seminar on a completely new topic because it comes with a free lunch can be fun, in addition to being interesting, allowing one to daydream, learn something new, and possibly network with a different crowd of people. Writing an outreach article can be fun, while also reusing some data that would be stored for nothing, and be a way to showcase your writing skills and outreach efforts, while feeling good and proud when it comes out. The examples are multiple on how to include joy in your work.

The flexibility in academia can also allow you to explore fun side projects that you are very curious about but have no funding for them because they too 'exotic', so-called 'un-scientific', and purely 'curiosity-driven'. You would be surprised at how undergraduate students are happy to take on any project you may propose, even if they are very unusual. While as assistant professor you want to minimize the risk of putting effort in a project with no result, you can arrange your research agenda to have non-risky projects along with slightly more risky ones.

If enthusiasm is momentarily not there for whatever reason, it can be artificially created too. For example, you may remind yourself to smile, look happy, or do yourself a small prep talk before a lecture, meeting, or event, when you feel tired and unmotivated.

PERSONAL SHARING

I wrote outreach articles in popular science magazines by proposing an article to that magazine. I always find it very rewarding to do these kinds of outreach as I enjoy writing and the idea that the research can go outside the lab. I also paid for a cover image for a scientific journal that in the end was not selected and wanted to reuse it somewhere although in the end, it still was not used. Writing in outreach magazines does not really increase impact or help recruit students, but it feels good, and then we can showcase it when talking to friends and closed ones who are outside academia. So, I find it almost as rewarding as publishing a paper in a peer-reviewed scientific journal with high impact factor.

KEY IMPORTANT POINTS

Having fun at work:

- Identify what you like the most in the job.
- Create opportunities to do what you like the most.
- Make sure you have fun.

8.4 Managing Social Media

Social media is nowadays present in everyday life, and has permeated the private and work spheres equally. It is therefore important to consider how social media can be used effectively. This subchapter also discusses how to manage social media to protect both working time and own privacy.

8.4.1 Benefits and Disadvantages of Social Media Presence

Social media presence is a new trend in academia, yet there is still very little training available on how to use if effectively. Social media also can be used for work or for private matters, and further blurs the barrier between work and life. Social media usage is also changing and evolving very quickly, with new platforms being created regularly and addressing certain populations with various types of media. Using social media can provide new opportunities but also be time-consuming and stressful, as it increases exposure to negative or critical comments.

Some positive aspects of using social media from a professional perspective is that it can help connect with other researchers and increase and diversify your professional network. Many conferences, journals, university, alumni groups, etc. are present on various social media platforms. Posting conference participation or recently published papers on social media can increase their impact and readership, leading to more citations, or create networking opportunities at conferences. Social media can also be used to advertise positions and help you find your next PhD student or researcher. Social media can also be used to find information and gain knowledge. Online platforms can give you access to recorded seminars and presentations by professors on various topics, including research, outreach and career.

Social media presence may, however, have some negative impacts, besides being time-consuming. Negative impacts can be to be receiving unnecessary and undesired comments or solicitations, for example by 'trolls' or other. This requires putting in place safe guards, being aware of the risks, and to not be exposing too much. Social media presence may also be frowned upon by others as being superficial and not professor-like. Acquiring adequate skills to follow social media trends can take energy.

Overall, social media presence can have a huge influence and benefits, but comes with some risks. Knowing about the various platforms, their audience, and the social media culture is helpful in deciding whether to engage on social media or not. Depending on your specific field of work, your colleagues may be using one platform more than another one. What to write and how to write on social media is also something to learn, to be effective quickly.

PERSONAL SHARING

Social media presence means that you can control what is said about you, especially how your profile appears in search engines like Google, Safari, Bing. Now, most of my professional social presence is through LinkedIn. Writing posts in LinkedIn can take some time. The development of chatbots like ChatGPT can be used to write social media posts quickly and may appear to save time. But posting using AI lacks authenticity and may downgrade your image, as it is obvious which posts were written using AI. Now I find it very hard to read these long posts with multiple fonts and symbols. I prefer short, concise, and authentic posts. People like social media because it feels like a closer connection, so my advice is to keep it authentic and natural.

KEY IMPORTANT POINTS

Benefits and disadvantages of social media presence:

- Social media presence can help your career in some ways.
- Social media presence also comes with some risks that need to be assessed.

8.4.2 Deciding to Engage on Social Media or Not

Before you decide to engage on social media professionally or not, you may want to think about the following questions to start with:

- Will you be posting on the social media as an individual, or for your research group?
- If it is for your research group, who will manage the social media?
- If the social media is for you as an individual, what will you post on it? Will it be purely professional or also personal?
- Who is your target audience?
- If you are posting as an individual, do you want your students to follow you? If they follow you, will you follow them back as well?
- Should the social media content be open to anyone or private and only for followers you have approved?
- How often do you intend to post?
- What is your preferred format? Will it be text, videos, links, images, a mixture of all?

- Do you have any specific goals for having a presence on social media?
- How will the social media benefit you?
- Why do you want to join and use social media?
- Which social media platform will you use? Will you use several platforms? If several platforms, will you post the same content on all platforms? Will they be connected?
- ...

PERSONAL SHARING

I started my own blog at the end of my PhD because I just simply wanted to write. I also found blogs by other faculty members to be eye-opening. Because I enjoyed reading other people's blogs and learning about their experiences, I also thought that if I write something, better have someone to read it. I also use blogging as a tool to frame my ideas as well as to improve my writing skills. I used to share my blog posts on other social media to get some clicks, but only for posts I felt were interesting enough to share.

KEY IMPORTANT POINTS

Deciding to engage on social media or not:

- Be prepared.
- Identify which platform, the audience, content and what you want to get from it and why.

8.4.3 Tips and Suggestions for Social Media Presence

The following provides some small tips and suggestions for social media presence. Because social media is always changing and evolving, you can allow yourself to experiment first. This means exploring several social media platforms to find the one that is the most interesting and useful, the one you like the most.

You can consider joining several social media platforms and separating each platform for different usage. For example, you can use one for your personal profile which you can keep private, while another one for advertising positions and your papers, and another one for keeping in touch with the research community and networking. One suggestion can be to avoid sharing too much personal or sensitive content on your professional social media and to keep the same level of professionalism as you have in everyday life.

To grow your social network, you can start by following colleagues, previous connections, people you know and have met in person, to avoid 'unwanted' connections. You should ideally develop a 6th sense for scams and trolls.

Be aware that social media can be highly political, as witnessed by the impact of social media on election outcomes and the sharing of fake news. You certainly do not want any misuse of what you are posting.

Universities may start writing policies about social media and provide training on social media presence. It is helpful to have a look at them if there is any. Your social media activities should represent your own views and not those of your institution. Keeping some records of selected posts on social media can be useful for showing impacts, or maintaining connections. You may download your data from the platforms or take screenshots.

PERSONAL SHARING

I use LinkedIn and Bluesky for professional sharing only, while I use Facebook and Instagram for personal sharing. I noticed that posting pictures always gives the most reactions on these four social media platforms, especially when there are people on the pictures. Each platform takes some time to know the rules, habits, and understand the culture of these platforms. So for professional posting, it is better to start with one, and learn how to work with it.

KEY IMPORTANT POINTS

Tips and suggestions for social media presence:

- Choose the right platform.
- Check policies.
- Stay professional.

8.5 Conclusions

Academic jobs are very demanding and it is necessary to create your own limits and expectations keeping in mind your personal and professional goals as well as what is expected from you. You can expect a lot of solicitations, hence it is key to start developing strategies early on how to protect your own time. Such strategies include learning to say no, being choosy, and making the most of academic flexibility. Short and long breaks should be incorporated into the turmoil of other activities and will come with huge benefits. Making sure you enjoy your work is capital. The work-life balance struggles on social media mirror in some ways the work-life struggles in real life. Developing a strategy to manage social media effectively will allow to make use of its capability.

Additional References

Additional References on Work-Life Balance in Academia:

E. Diego-Medrano, L.R. Salazar, Examining the work-life balance of faculty in higher education, Int J Social Policy Educ 3 (2021), 27–36.

Additional References on Work-Life Balance in Asia:

N.S. Jaharuddin, L.N. Zainol, The impact of work-life balance on job engagement and turnover intention, South East Asian J Manag13 (2019), 1–13.

H. Le, A. Newman, J. Menzies, C. Zheng, J. Fermelis, Work-life balance in Asia: a systematic review, Hum Resour Manag Rev 30 (2020), 100766.

Additional References on Social Media Presence:

C. Machado, P.Y. Hdia, C. Vaccaro, C. Baker, Entering the social media stratosphere: higher education faculty use of social media with students across four disciplines, J Inter Media Educ 1 (2024), 1–17.

Y. Chen, H. Rui, A.B. Whinston, Tweet to the top? Social media personal branding and career outcomes, MIS Quarterly 45 (2021), 499–534.

R. Chugh, R. Grose, S.A. Macht, Social media usage by higher education academics: a scoping review of the literature, Educ Informat Technol 26 (2021), 983–999.

M.J. Matthews, S.H. Matthews, D. Wang, T.K. Kelemen, Tweet, like, subscribe! Understanding leadership through social media use, Leadership Quart 33 (2022), 101580.

https://www.imperial.ac.uk/staff/tools-and-reference/web-guide/tools/social-media/

9

Ensuring Diversity

9.1 We Need Diversity

> *Diversity in gender, age, background, ability, etc. is increasingly stressed as very important in universities and other workplaces. This subchapter further highlights how diversity is also important at the level of an assistant professor, in particular in the research group, and how this diversity can serve the research and its impacts.*

Recent years have witnessed an increasing awareness about the lack of diversity in certain spheres of the society, and the need for more diversity overall. This is particularly true in universities, especially when going higher in the academic ranks. This is also particularly true in STEM departments. Although most of the studies on the topic come from Western countries, the lack of diversity is a global phenomenon, which nevertheless is also affected by local culture and local context. Furthermore, diversity refers to diversity in people's background, nationality, gender, etc. While diversity is important, creating a diverse environment does not necessarily lead to inclusivity. Leadership with a diversity and inclusivity mindset is necessary to let the positive impacts of diversity blossom. Reported positive impacts of diversity include more publications, more novel and impactful research, and increased overall team performance and engagement, etc.

From an assistant professor's standpoint, diversity can be achieved when recruiting students and staff. Diversity can help in getting visibility and recognition. When you have a diverse and inclusive research group that provides a comfortable and inclusive working environment, students and researchers may talk about it within their circles and therefore help you with the recruitment. In a research group, diversity can be efficiently combined with multidisciplinarity, as certain groups of people tend to study certain topics. Hence, diversity is also helpful for research and innovation, especially when the funding landscape calls for more multidisciplinarity.

The idea of having multidisciplinary projects and a diverse team can be implemented by considering research projects in a more holistic way, which could increase the impact of the research. If the impact may not be directly magnified by multidisciplinary and diverse teams, it will likely be broader as

DOI: 10.1201/9781003635116-9

publications and other forms of communications may reach various scientific or other communities. Another benefit of multidisciplinary and diverse teams can be to look at a problem from a different or unusual angle, and therefore to provide another innovative solution. Overall, when you talk to people working with a multidisciplinary and diverse team, you will hear that this makes the research more fun because there is a strong learning component that goes beyond the specific scientific or technical fields. The human aspect that comes with such teams is a great source of personal enjoyment and growth.

One also needs to be aware that diversity is becoming a necessity for journals and conferences, although still applied inconsistently. Journals and publishers keep records, and some personal data is often requested when submitting a paper or agreeing to peer-review. Diversity in conference lists of keynote and invited speakers, chairs, etc. is also increasingly taken into account, and it is a good idea to therefore have a diverse pool of colleagues. Finally, diversity criteria may also now be necessary to apply for certain grants.

PERSONAL SHARING

I have seen colleagues refusing invitations to participate in a conference because the keynote speaker list did not reflect diversity, in that case it was gender-diversity. Diversity does not occur by magic, it should be carefully thought of and intentionally crafted. I am in a society committee where we organize international conferences. The president of that society makes it a requirement to have diverse pools of keynote and invited speakers. In particular, we have to pay attention that they come from different regions of the world, different nationalities and ethnicities, and that we have gender-balance. I have seen this at a couple of other conferences too, but not at every conference.

KEY IMPORTANT POINTS

We need diversity:

- Diversity, with inclusivity, has many positive outcomes.
- Diversity can be a prerequisite in job applications, grants, etc.

9.2 Practical Tips to Create a Diverse Team

Practical tips to trigger or ensure diversity in the workplace, the university, the department, and in the research group are given in this subchapter. From the standpoint of an assistant professor, it is important to not only foster diversity in your own team, but also to contribute to the diversity of the university at large.

9.2.1 Diversity during Recruitment

Creating a diverse team should be crafted intentionally at each stage of the recruitment for your team. To create a more diverse team, whatever adjective you put under 'diverse', one first action is to make a statement. This statement can be written on your website or can be written on job applications you advertise for your research team. Reading a statement can encourage people from that diversity to apply, as well as people who have the same mindset. The job advertisement should also be written to reflect that diversity, for example using gender-neutral terms, or presenting the topic without assuming specific knowledge from the applicant, for example without using very specific terms for a very specific research topic. If you are looking for a specific type of profile to enrich your team, you can also scope down how that profile should be.

The job advertisement should be sent out also with your diversity goals in mind. If you want to recruit diverse nationalities, you can talk to foreign students in your university to share your job advertisement, or advertise to colleagues in a foreign country/university. If you are looking for more women, you can ask women colleagues and students to share and talk about it, etc. Diversity can be sought out proactively. You can also leverage your international network of collaborators and colleagues to find applicants from other countries or with different types of background.

Then, when reviewing the applications, you can also take the right steps to make sure you have the diversity that you want. A simple way is to mention clearly that you want an international, multidisciplinary, inclusive, etc. team. You should also be aware of your own unconscious biases.

PERSONAL SHARING

I find that recruiting diverse students from different countries or women PhD students to be challenging. Most applicants to my PhD student positions come from nearby countries and are men with training in mechanical engineering. I found a great way to at least bring attention to your own opportunities is through the students themselves. If they enjoy their research, they likely will share with their friends and tell them to apply as well. My existing international network did help me recruit one PhD student. In general, I find undergraduate and postgraduate students are more likely to come from nearby countries. Hence, international diversity arises more at the researcher level, or for short-term exchanges, where tapping into alumni and existing exchange schemes at the university are helpful. Diversity in expertise can be more easily set by scoping different projects and using keywords. Gender diversity is also difficult to have, but I found that just me being there helps in having more female applicants than for other male colleagues in mechanical engineering.

KEY IMPORTANT POINTS

Diversity during recruitment:

- Make a statement.
- Be aware of your own biases.

9.2.2 Create an Inclusive Atmosphere

Recruiting for diversity is one first step. Then, you also want to retain this diversity and make this diversity productive. There is therefore the need to ensure that everyone is comfortable. Although it may look like it is not possible to make everybody happy, it is totally possible to create a group with a supportive and inclusive atmosphere. Likely, you can tap first into the university facilities that can provide some support and guidance if necessary. Within your research group and how you manage it, you can also take some actions to make it a safe and inclusive environment for the benefit of all, including your own. This is referred to in the literature as inclusive leadership.

Simple steps to create an inclusive environment are:

- Make people aware of the diversity. Without pointing out or singling out individuals if it is a sensitive type of diversity, you can keep statistics and share them with your team (Figure 9.1). Sharing this information can reinforce the emphasis that you place on diversity and will incite your team members to follow and be proud of it. It will create a diversity awareness and mindset among your team.
- Use inclusive language: Being inclusive can mean using gender-neutral terms, but also taking the time to explain things that may be obvious for one population and not so for another. This can be explaining specific scientific or cultural terms without being condescending. Inclusive language can also include welcoming and encouraging questions, which can be easily demonstrated by yourself asking questions.
- Have inclusive considerations: You can consider celebrating major cultural events or cultural holidays with a small note, and pay attention to various cultures. When a group of people feel comfortable, each person will want to celebrate what they feel like sharing and celebrating. This can become team-bonding events and be opportunities for learning (and celebrating).
- Get feedback and push a little: Sometimes, asking people what they need is a clear and straightforward way to ensure the workplace is suitable. You can share that you are open for feedback, which your team members can provide later in private meetings or by email when they feel the need to.

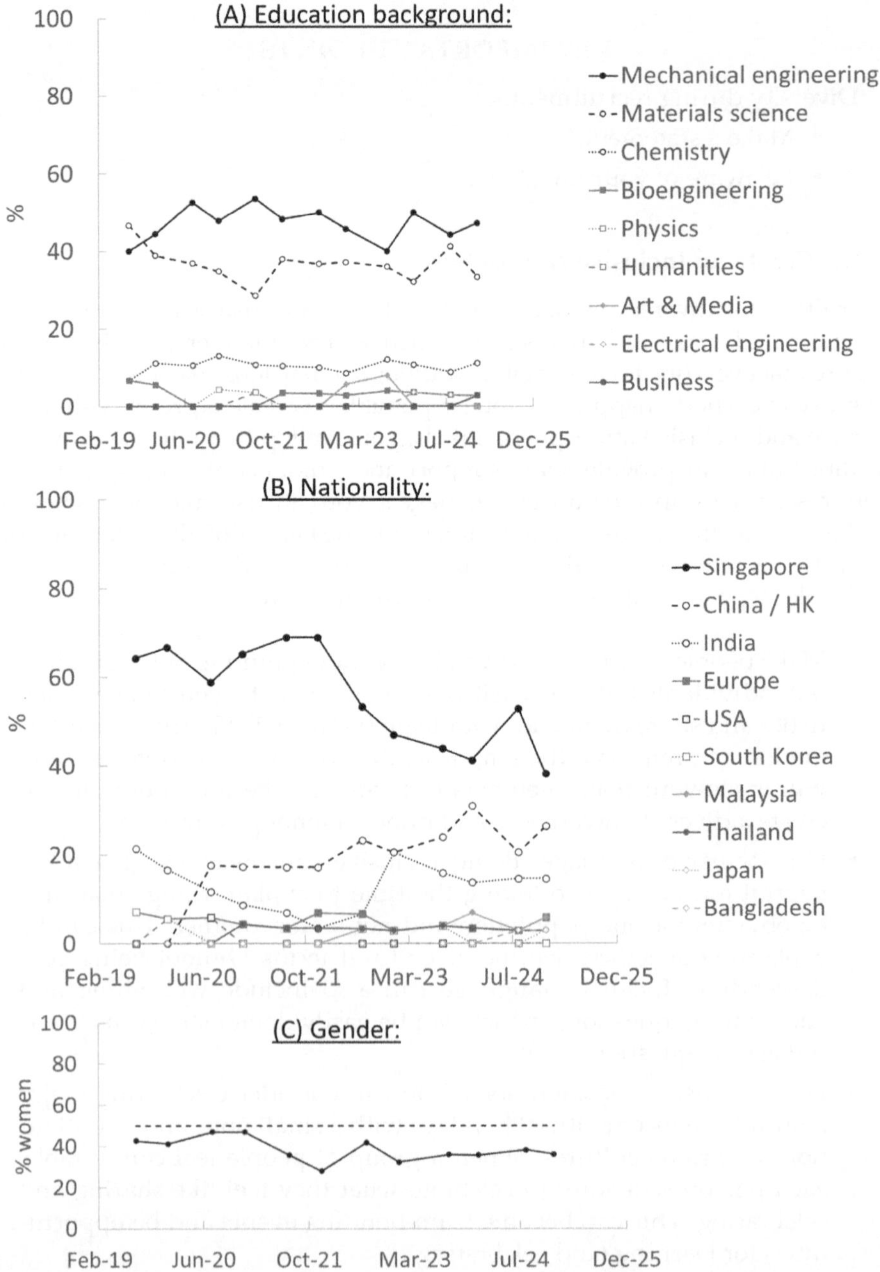

FIGURE 9.1
Example of statistics collected to reflect diversity: (a) education background, (b) nationality, (c) gender.

- Have a clear structure: It is helpful for people to adapt to their own needs when there is a clear structure to work with. Having regularity in your interactions and some guidelines will help people to adjust around it. Academia is known for being a stressful place where students and researchers work very hard. Having a safe and regular planning of meetings and discussion, safe and comfortable meeting rooms, etc. can be calming and relaxing for everybody, providing a supportive environment.

PERSONAL SHARING

I had mentored and supervised students who were neurodivergent. Reading and getting to know more about what that topic in the literature helped me to create an environment to support them as best as I could (definitively better than without reading about it). While we receive no training about leadership and even less about inclusive leadership, self-studying can be game-changing. The strategies I implemented to support those students proved to be beneficial for the entire group as well as for myself. These include setting a clear structure and organization for my research group, with resources available on the internal website, multiple channels of communication, a survival guide, and me being more aware and knowledgeable about how the university works. I am grateful to those students although I could still do better.

KEY IMPORTANT POINTS

Create an inclusive atmosphere:

- Make people aware, get feedback, ask questions and be adaptive.
- Create a support structure.

9.2.3 Reflect on Your Own & Others' Experience

While considering how to create a diverse research team, you can also reflect on your own experience and your own encounter of diversity or absence of diversity:

- How is the team that you would like to have?
- What is in your capacity to create and ensure a safe and comfortable place for your team members?

- What are your own limits in the support that you can provide?
- Does your university have a well-being centre or other sources of support?
- How do your colleagues deal with different types of profiles? What is the diversity culture in your department and what do you think about it?
- What is the diversity in the pool of students you have access to and how does it match with what you would like to have?
- Do you have any other example of diverse research team you can take inspiration from?
- What did you like, and did you not like in the research groups you were part of previously? How would you do things differently?

Ideally, you would like to create a supportive and inclusive research team from the start, that can sustain itself continuously afterwards. This is why it can save time and efforts to learn and think about it early. It can also be helpful to read broadly on the diversity of profiles of students that you may have and familiarize yourselves with various cultures or conditions to be able to tackle them sensibly.

PERSONAL SHARING

I have been in research groups where nobody talked to me for months. That felt very odd to me. People did not really know who was doing what and only met during meetings. I have also been in research groups where several people were taking care of you the first few days, showing you around, taking you to lunch, etc. In these groups, people sat together, chatted, and there was a lot of life. Because I preferred the latter experience, I arrange to link any new member with others in my team and emphasize that I want people to help each other and that the overall atmosphere is important to me. I also say I encourage collaboration, leading to benefits such as being co-authors of each other's studies if significant contributions can be made. Since then, many collaborations have sparked within the group. Also, there is a good atmosphere and a strong bonding which I think is very pleasant. I was particularly surprised and happy to learn that the lab manager also enjoys the group and that they have their own party events. Having a good relationship with lab managers is good because they oversee the safety of the lab.

KEY IMPORTANT POINTS

Reflect on your own & other experience:

- Identify what is the ideal diverse team you would to have.
- Look for examples or counter-examples.
- Learn about any university support.

9.3 What to Do If You are Part of 'Diversity'

This section discusses challenges that you may face if you are part of that diversity and provides suggestions on how to address them.

9.3.1 Make Sure You Belong

If you are part of the 'diversity' and are or feel like you are part of a minority in your area, it is capital that you feel like you belong where you are. While you would expect that it is the role of the university to provide you with this feeling, it may unfortunately not always be the case. To ensure that you feel you belong at 100%, you can take small proactive steps, such as looking for a supportive group from the same minority. This can be a few colleagues or even students, anyone who could listen and help if a challenge arises. Universities can be quite big and likely you can find other people to bond with. You can also find support in other nearby universities or organizations. If you want more support or more diversity, you can also invite people you know to come and present in workshops or symposia. It is always good to point out that these events will benefit a broader spectrum of people.

As part of the support, you may want to look for extra role models and mentors. Although likely everybody is supportive of you (remember that you have been hired because the university wants you to succeed), extra understanding and motivation can sometimes help when challenges arise. Role models and helpful suggestions can be found online easily when it feels difficult to find or to approach locally.

If you feel there are some inappropriate behaviours, or things that could be changed, you can check out your university policies and voice them out in a subtle way. The best way to make things change is to provide easily implantable solutions at the same time as raising a concern. It may also be good to talk about it privately with someone who can help to avoid misunderstandings. However, it is good to be patient as universities are complex and large institutions.

PERSONAL SHARING

I have been told at a conference by a colleague that 'I liked organizing symposia and workshops' and conducting activities for women in STEM. I do not particularly like organizing events in general. It takes time, coordination, and effort. However, what I like is to listen to my friends and to other inspiring speakers that I cannot always find in my own direct environment. I have been organizing networking breakfasts for women in STEM at the university to meet and discuss. That was a nice opportunity to learn from each other's experiences. I have organized symposia and workshop on biomimicry and bioinspiration because this is one of my big motivations and research inspiration. I have been inviting science communicators to share their experience and tips because I strongly believe we need to improve our communication skills, especially for broader outreach.

KEY IMPORTANT POINTS

Make sure you belong:

- Find your supportive group.
- Create your supportive group if you cannot find it.

9.3.2 Embrace Your 'Diversity' Profile

Although you do not want to be recognized solely because of your 'diversity' profile, you can use it to bring extra value to your team, your colleagues and university, and to your research. For example, you can join recruitment committees to contribute with service to your school and bring in more diversity (make sure to select carefully which committees you agree to join). You can point to diversity as a key advantage for the reputation of your school.

You can also apply to special grants, fellowships, and awards that may exist for your profile. You can find out about those special grants online or by asking some AI tools like ChatGPT. Using keywords related to your specific profile likely will lead you to finding some opportunities. In addition to the reward and personal pride you get when winning these, they will add a plus to your tenure dossier as being unique compared to other more conventional profiles. It is therefore even better if you can connect any diversity aspect to your research in some ways. Linking any extra activity to your research can be an add-on to your CV under the outreach section and therefore be another strength to add to your tenure dossier.

Finally, you can also be an advocate to the students and be a role model yourself. You can volunteer to participate in panel discussions or interviews to share your experience. It should not feel strange or uncomfortable to volunteer for these activities if they are meaningful to you and can provide support for others. Again, be extra careful to not add too much to your plate and therefore select what you commit to. It can easily become a lot as people who have talked in some panel will likely be invited to other panels.

PERSONAL SHARING

Applying and getting awards for women in STEM clearly was a big addition to my profile as well as a great boost in confidence and further motivation to go on. I have also been congratulated several times when I was volunteering for interviews or simply sharing my views. I felt it was minimum effort with maximum benefits. Just by being out there, I have had students, parents, even working professionals congratulating me. This was surprising at first, but definitely very empowering. For the tenure-track perspective, women-specific awards have the same rigorous evaluation and high standing as any other award, and often come with multiple networking and mentoring opportunities. A common challenge for awards is that there are not enough applications, so being a proactive applicant is definitively rewarding.

KEY IMPORTANT POINTS

Take advantage of your 'diversity' profile:

- Showcase your diversity.
- Showcase your uniqueness.

9.4 Conclusions

Diversity can be a buzz word and encompasses many different meanings. Defining the diversity that is meaningful to you can help you in having a unique research profile and research team. Creating and ensuring diversity in your team has multiple benefits and needs to be crafted and taken care of intentionally. This starts from the recruitment, the everyday atmosphere, and support. You can use your own experience as well as the literature to find concrete approaches to implement this diversity effectively. If you consider you are yourself part of diversity, you should also take proactive steps to ensure you feel you belong, by finding or creating your own safe space. You can also embrace your unique profile to enrich your research and CV.

Additional References

Additional references about diversity in academia:

A.C. Morgan, N. LaBerge, D.B. Larremore, M. Galesic, J.E. Brand, A. Clauset, Socioeconomic roots of academic faculty, Nat Human Behav 6 (2022), 1625–1633.

J.D. Llamas, K. Nguyen, A. Tran, The case for greater faculty diversity: examining the educational impacts of student-faculty racial/ethnic match, Race Ethn Educ 24 (2021), 375–391.

C.H. Dupree, C.M. Boykin, Racial inequality in academia: systemic origins, modern challenges, and policy recommendations, Policy Insights Behavioral Brain Sci 8 (2021), 11–18.

Additional references about inclusive leadership:

K.M. Rosenkranz, T.K. Arora, P.M. Termuhlen, S.C. Stain, S. Misra, D. Dent, V. Nfonsam, Diversity, equity and inclusion in medicine: why it matters and how do we achieve it? J Surg Educ 78 (2020), 1931.

S.C.L. Kamerlin, When we increase diversity in academia, we all win, EMBO Rep 21 (2020), e51994.

M.A. Fuentes, D.G. Zelaya, J.W. Madsen, Considerations for promoting equity, diversity, and inclusion, Teach Psychol 48 (2021), 69–79.

Y. Yang, T.Y. Tian, T.K. Woodruff, B.F. Jones, B. Uzzi, Gender-diverse teams produce more novel and higher-impact scientific ideas, PNAS 119 (2022), e2200841119.

T. Ashikali, S. Groeneveld, B. Kuipers, The role of inclusive leadership in supporting an inclusive climate in diverse public sector teams, Rev Public Pers Adm 41 (2021), 497–519.

10

The Tenure Track Dossier

10.1 Preparation

The tenure clock starts on day 1 of your position as assistant professor on a tenure track. Understanding the expectations to obtain tenure is key. You can also set your own expectations and get started as early as possible.

10.1.1 The Expectations

To be able to prepare your tenure dossier, it is a must to know what is expected from you by your university and more specifically by your department. A typical tenure package includes a CV and a research, a teaching, and a service statement where you show your accomplishments. There may likely be other documents required, such as a list of potential reviewers, documents supporting the details that are presented in your CV, and maybe another benchmark table comparing yourself to others in the field worldwide, for example. How to get this information if it is not provided to you? You should ask your department Chair about it.

It is also imperative for you to know what the process for applying is, which means what the requirements are, before considering putting up your dossier. Requirements can be related to how many hours of teaching or the size of the teaching class, having graduated a student, etc. There may be additional reports for you to get, which you should be aware of early. This information you can request from your department Chair again if not provided directly to you. Knowing the process includes knowing about the timeline, which is when you can submit your dossier, how long it will take, whether there will be other documents or feedback after your submission, etc.

It can also be helpful for you to know about the success rate and to get informal tips about the entire process. There may be 'hidden' knowledge that can be helpful to get. A great way to get information is to talk to people who have obtained their tenure fairly recently, within 1–2 years. They will likely remember more clearly and be updated on the 'rules'. You should be aware that there may be university changes in the rules and evaluation of tenure dossiers.

DOI: 10.1201/9781003635116-10

PERSONAL SHARING

As I was in a joint appointment between two schools, I had to show how I fitted well in both schools. More specifically, my appointment was 75% in one of the schools and therefore I should make sure I fit well with that school. Because the 2 schools have different standards and metrics, I had to be a bit careful. I received feedback from senior faculty on what I should do to 'fit' in. In that particular case, it was through technology disclosures and patents, as that school values innovations the most. Getting to know about this early was definitively very helpful.

KEY IMPORTANT POINTS

The expectations:

- Get to know what is expected from your department Chair.
- Get to know what the process for applying is and what the timeline is.
- Get informal tips from trusted colleagues.

10.1.2 Setting Your Own Expectations

While there are expectations from the university for your tenure, you should also set your own expectations, which means you should define what you aim to achieve for your research, teaching, and service. This should be your major drive and you can set your own goals and expectations to be aligned with those of the university.

One way to visualize these expectations is through your short-term and long-term research, teaching and service goals, which you should take some time to reflect upon. These goals should then be made more specific so you know exactly what you are aiming at.

To help you craft these goals, you can input them in a 'Career table' as shown below, which you can re-evaluate every year or half-year. Seeing your trajectory will help you get everything you need for tenure. You can input your specific goals in the table, then edit what you have achieved and what you have not achieved. For example, you can strike through what did not work out so you can still see it in the table as an unachieved goal, or use a colour code (Table 10.1).

TABLE 10.1

Example of a Career Table Inspired from My Own

	Year	Year 1	Year 2	Year 3	Year 4	Year 5	Year 6
	General goal or achievement	Set up the laboratory		PhD student X and Y confirmation. Apply to award 1	PhD student Z confirmation. Apply to award 2	PhD student X and Y graduation. PhD student Z confirmation. Apply award 3. Submit tenure	PhD student Z graduation. Obtain tenure
	Specific requirement for tenure		Get grants. Publish at least 4 papers.	Get good teaching. Publish at least 4 papers.	Get better teaching. Apply for more grants. Publish at least 4 papers	Get very good teaching. Publications in top journals	
	Strategic planning	Finishing last publication from PhD/postdoc. Start getting students. Get funding	Recruit postdocs. Recruit master and visiting students. Get equipment …. Get new PhD students. Find more laboratory space.	Get new equipment …. Start new project on ….	Finish first grant. Higher more postdocs to work on ….	Do grant variation. New grant starts.	Applied to grant 1.
Teaching	Classes	Class … Tutorial …	Class … Tutorial …	Class … Tutorial …	Class … Tutorial …	Class … Tutorial …	Class … Tutorial …
	Mentoring	Nbr of PhD students, bachelor students, visiting students. Postdocs	Nbr of PhD students, bachelor students, visiting students. Postdocs	Nbr of PhD students, bachelor students, visiting students. Postdocs	Nbr of PhD students, bachelor students, visiting students. Postdocs	Nbr of PhD students, bachelor students, visiting students. Postdocs	Nbr of PhD students, bachelor students, visiting students. Postdocs

(Continued)

TABLE 10.1 (*Continued*)

Example of a Career Table Inspired from My Own

	Year	Year 1	Year 2	Year 3	Year 4	Year 5	Year 6
Research	Publications	Paper 1 (journal 1) Paper 2 (journal 2) Paper 3 (journal 3) Paper 4 (journal 4)	Paper 1 (journal 1) Paper 2 (journal 2) Paper 3 (journal 3) Paper 4 (journal 4)	Paper 1 (journal 1) Paper 2 (journal 2) Paper 3 (journal 3) Paper 4 (journal 4)	Paper 1 (journal 1) Paper 2 (journal 2) Paper 3 (journal 3) Paper 4 (journal 4)	Paper 1 (journal 1) Paper 2 (journal 2) Paper 3 (journal 3) Paper 4 (journal 4)	Paper 1 (journal 1) Paper 2 (journal 2) Paper 3 (journal 3) Paper 4 (journal 4)
	Conferences	Conference 1. Conference 2. Conference 3.	Conference 1. Conference 2. Conference 3.	Conference 1. Conference 2. Conference 3.	Conference 1. Conference 2. Conference 3.	Conference 1. Conference 2. Conference 3.	Conference 1. Conference 2. Conference 3.
	Grants	Apply grant 1. Obtain grant 2.	Apply grant 1. Obtain grant 2.	Apply grant 1. Obtain grant 2.	Apply grant 1. Obtain grant 2.	Apply grant 1. Obtain grant 2.	Apply grant 1. Obtain grant 2.
	Industry		Sign research collaboration with industry 1	Technical disclosure 1.	Technical disclosure 1.	Consultation with company 2. Technical disclosure 1.	
Service	Committees	Committee 1.	Editorial board 2	Committee 1. Editorial board 2	Committee 1. Editorial board 2	Committee 1. Editorial board 2	Committee 1. Editorial board 2
	Other service	Judge event 1.	Reviewer.	Symposium organization.	Reviewing grant.	Judging.	Reviewing.

The number of journals, students, committees, etc. are to be set by you and are non-indicative here

PERSONAL SHARING

One way to find your own expectations can be by looking at the CV or accomplishments of other colleagues you admire or who are a few years ahead of you. Such information can be found from their website or social media and give you more concrete specific targets to aim for, such as awards to apply to, number of publications, participation in committees, etc. For the tenure evaluation, you may be asked to compare yourself and your achievements with those of others internationally. Hence, having a look at what other people do will help you set goals to perform better.

KEY IMPORTANT POINTS

Setting your own expectations:

- Align the expectations from your department with your own goals and aims.
- Have a clear long-term vision for your tenure track with very specific targets.

10.1.3 Getting the Process Started

How to get the tenure process started? Once you join as an assistant professor on a tenure-track, the tenure clock starts. Once you have clarity about when you are planning to be ready according to your own expectations, you should arrange a talk with your department Chair to inform them of your plan. They may give you feedback. Likely, if they feel you are not ready, or that your plan needs to be adjusted, they will let you know. There is no interest for a department to put someone forward for tenure if they do not support them fully. They want to make sure that the applicants for tenure will be successful.

After you have decided when to apply, make sure you get all the templates required to prepare your tenure dossier and plan ahead on when to fill them up. Because there may be strict deadlines on when you are to submit the tenure dossier, you may want to be ready by these deadlines.

Pay also attention that you will be asked to name reviewers who can read your tenure dossier and provide comments on your promotion to the next level. Finding people to name can take some time. It requires connecting and networking so that people know about you and about your work, and have a favourable impression of you as well. These people should ideally be full professors and be well known in their field and be employed at reputable universities. While you will put up a list of potential reviewers, you do not know if they will be selected as reviewers or not, whether they will accept to do it because of time or other reasons. It is therefore a good idea to start early, growing your own network and your own visibility in your scientific community, so you can have a list of people ready to be presented.

PERSONAL SHARING

One challenge is often to define when we want to be put up for tenure. Universities may have different clocks and may allow being put up a second time if the first application is not successful. In my case, there were two specific requirements to be put up for tenure, which were having graduated a PhD student and having taught a large class. I had decided myself from the start that I would apply for the tenure after my first PhD student graduated, leaving me with a few more years to reapply in case I was not successful. I also got to know that in addition to all the papers and conferences, having at least one ongoing grant is important, in order to show that you can 'survive' after tenure. I managed to get good publications the year before, to get awards as well, and I met all these conditions, so I felt confident about going for tenure.

KEY IMPORTANT POINTS

Get the process started:

- You decide and plan when you want to be promoted.
- Start from day 1 to build on your tenure package.

10.2 The Tenure Dossier

The tenure dossier contains the overview of your work since you are on the tenure-track, as well as what you will deliver afterwards. Consistent and organised documentation of your work is needed. Then, you will have to write your research, teaching, service, and future aspiration statements.

10.2.1 Documenting Your Work

To be able to write your statements after 4–5 years of employment, and to have the documents supporting what you have listed in your CV, it is critical that you start documenting your work from day 1. Documenting your work means saving the evidence of every achievement or other work activities that are relevant to your tenure. You can take screenshots of websites where your name is mentioned or where your work is cited, which can include social media or blogs. You can take pictures of cards or letters that your students give you. You can save emails with invitations to present your work, invitations to join a committee, to review a grant, etc. Save these documents into well-organized folders and sort them by the date. You can save them as they appear and sort them out later. This will save you precious time when they will be needed. Particularly, save anything you think is relevant to your own expectations. These documents will serve as evidence when you will put forward your application. Everything you do, indeed, should go either in your CV or in your statements, somehow.

As a guide, below is a list of folders that you may want to create to start sorting out these pieces of evidence.

Main folder 'Research':

Sub-folders:

- Awards & Recognition
- Citation counts
- Conference presentations
- Exhibitions
- Grants
- Industrial collaborations & consultations
- Media reports
- Patents & Technology disclosures
- Publications
- Seminar & webinar presentations
- Start Ups & Entrepreneurship
- Research Outreach

Main folder 'Teaching':

Sub-folders:

- Course taught with number of students
- Course evaluation
- Example of innovative teaching approaches
- Laudatory notes & emails from students
- Mentorship talks
- Mentorship outreach
- PhD and master theses committees
- Roles in curriculum changes and development
- Student achievements
- Papers co-authored by students

Main folder 'Service':
Sub-folders:

- Conference organization
- Editorial roles
- Judging roles
- Leadership roles
- Letters of recommendation
- Memberships
- Peer reviewing of grants or papers
- School service

PERSONAL SHARING

Some conference websites disappear from the internet after a few years, which makes it difficult to provide proof for invited talks, for example. I recommend doing record-keeping regularly, even if it is not useful in the end. I take screenshots of everything and as soon as I have a paper published, a new student starting a project, or an invitation to talk at a conference, I add it to my CV so I am sure not to forget about it. At the same time, I add it to the group website so that I have another place where I know what I have done.

KEY IMPORTANT POINTS

Documenting your work:

- Practice regular and comprehensive record-keeping.

10.2.2 The Research Statement

The research statement should clearly give the vision of your research, with the short- and long-term objectives. You should spell out all your achievements, mentioning all your papers and how they contribute to your research objectives. You should clearly indicate the novelty and impact of your work. You can provide tables and data to support your statements, for example, with a table containing the titles of the journals you published in, to indicate how your research is relevant to one specific field, or several fields. You can also make a table to list your outreach activities if you have been for example, conducting workshops or participating in exhibitions to showcase your research.

Below I provide an example of a structure for your research statement:

Introduction

Your research statement can start with an introduction pointing out what your research is about what its global impact is. You can refer to your affiliations if they contribute to supporting your research. Then, you give a brief presentation of what your statement contains.

Summary and context of the research.

In this section, you can explain the specific research aims that you are looking into in order to achieve your major global research aim. If you have published reviews or perspectives that support your vision, you can mention them. Here is a good place for a table with your top-tier peer-reviewed journal publications since you joined the faculty.

Specific scholarly contributions

Here you can give more details about your research. You can list the research questions that you are trying to address and how you are tackling them. For each research question, you can mention your key contributions and the papers that you have published. You can also point out any novelty in terms of equipment or approach and acknowledge key collaborators.

Research applications and impact

This section aims at explaining what the impacts of your research are. There are three types of impacts. First are the impacts on the scientific community. There you can highlight your citations, your invitations to give talks at international conferences, or to give seminars. You can mention collaborations and symposia organization. Any paper highly cited or appearing in Google Scholar first upon a key search should be highlighted, as well as media reports and awards. The second type of impact is the impact on your region or country. This can be the opportunity to mention industries or start-ups collaborations, media releases in regional or national newspapers, or other local news media. The third type of impact is impact on the society where you can mention more global outreach and media release, international collaborations or participation in events showcasing your work. You can report collaborations or teaching in other fields. To provide evidence of media reports, you can write down the URLs that link to the respective websites.

PERSONAL SHARING

I spent a lot of time trying to define my research in one sentence because I had several projects on slightly different topics. Then, I divided my research into four pillars, which are not necessarily related to four different grants. Having the PPT as described in Chapter 2, Section 9, was instrumental for me to creating my storyline.

KEY IMPORTANT POINTS

The research statement:

- The statement builds a story about what you have accomplished in your research, highlighting its impacts.

10.2.3 The Teaching Statement

The teaching statement should start with your teaching philosophy and comprises your teaching in classes, tutorials, and laboratories, but also mentorship of students including postgraduate students.

Below I provide an example of a structure for your teaching statement:

Introduction

You can start by explaining what type of teaching you have been involved in, in which department, and how the teaching is conducted. Then you introduce the scope of the statement.

Summary and teaching philosophy

This section explains what you aim to achieve in terms of teaching and what your goals are. You can list down the topics you have been teaching and the challenges faced when teaching these topics. You should explain your teaching approach to address those challenges and what the benefits of your approach are to the students.

Pedagogical innovations

You can provide more details about your teaching approach. Ideally, you have several methods or strategies to achieve your teaching aims. You can explain why you think these strategies will be efficient and how you came up with them, leading to details about your methods and the results. In terms of results, you can quote comments left by the students in your classes. If you can support your teaching approach with references from the literature, this will be even more effective and will position you as a thoughtful educator. Any data that can support your statements can also be provided, which may be statistics of your student population or providing illustrations of your methods.

Key student's achievements

In this section you can list the achievements of the students that you have mentored to show your skills as a supervisor and mentor. If one of your students has won a prize or an award, it can be mentioned there. You can also list the papers published by undergraduate students or if they have pursued their studies in a specific field or university that you know of.

PERSONAL SHARING

Same as for the research statement, the teaching statement should have a storyline that includes all your teaching activities, as diverse as they might be. I wrote a teaching philosophy sentence as a pitch to what a good educator should be in my view, then divided it into three main points, which I detailed, where each point related to one of my teaching approaches.

KEY IMPORTANT POINTS

The teaching statement:

- The statement should describe your teaching philosophy and your teaching activities and their impacts.

10.2.4 The Service Statement

The service statement will outline your service activities and the motivations for doing them. You want to explain what these services are and their impacts.

Below I provide an example of a structure for your service statement:

Introduction

You can start by giving an overview of your service contributions.

Service to the university

Here you detail your service to the university, which may be to mentor students or student groups, be an advisor or an examiner of students' theses, be a judge at events, or undertake any other activities with university students. You should explain why you took these activities and what your aims were. If possible, any achievements should also be mentioned. Depending on the activities, you can prepare tables to list them. For example, if you had been examiner of multiple PhD or master's theses, you can make a list.

Service to the scientific community

Here you can also detail reviewing activities for journals and grants, and participation to the organization of symposia and conferences. You can also list your editorial roles or other memberships.

Service to the society

Here you can detail any broader service such as outreach and science communication activities. You may report workshops for pre-university students or any member of the society, blogs, social media activity, videos, or interviews you have given.

Examples of tables to support service statements (Tables 10.2–10.7):

TABLE 10.2

PhD and Master Theses You Have Been an Advisor and/or an Examiner

Number	Student Name	Degree	School	Year	Role

TABLE 10.3

Summary of the PhD Theses You Were the Jury Outside Your Own University

Number	Title of the Thesis	Name of the Student	University and Ranking	Country	Role and Year

TABLE 10.4

Overview of Your Journal Peer Review Activities Since You Joined Faculty

Focus of the Journal	Journal Names and Impact Factors

TABLE 10.5

Summary of the International Grants You Were a Peer Reviewer

Number	Funding Agency	Name of the Grant	Country	Description of the Grant	Link

TABLE 10.6

Summary of Your Roles in Conference Organization Since You Joined Faculty

Number	Title of the Conference	Role	Scale of the Event	Ranking and Importance	Website

TABLE 10.7

Summary of Your Editorial Roles Since You Joined Faculty

Number	Journal Title	Role	Dates	Ranking and Importance	Website

PERSONAL SHARING

Any other activity I had done during the tenure-track that did not fit in the other 2 statements found its way into the service statement, such as judging events, media, or other outreach activities, in addition to reviewing and participation in committees. I find that national and international services are viewed more positively, although other services also have their value when presented appropriately. Each service should be introduced with a specific purpose or impact.

KEY IMPORTANT POINTS

The service statement:

- The statement will cover all service activities to the university, the scientific community, and the society.

10.2.5 Future Aspirations Statement

The future aspirations statement is optional and can be merged in part with the other statements. The aim of this statement is to explain what your plans are in terms of research, teaching, and service for the coming years after tenure. It is the opportunity to spell out on-going work you have been doing that is still preliminary, such as new side projects, new collaborations, and new approaches. Since you have a short- and a long-term research plans, you can describe the on-going steps to realise them. For simplicity, you can separate your future aspirations for research, teaching, and service into 3 small paragraphs.

PERSONAL SHARING

My future aspiration statement was only 2 pages with one large figure illustrating my research vision for the next 5–10 years. This was much shorter than my other statements, which were approximately 4–6 pages long. While I presented what I want to do in the future, I realize I am already going in another direction.

KEY IMPORTANT POINTS

Future aspirations statement:

- This statement is to describe what you will accomplish after you get tenure.

10.2.6 Set the Tone

These statements should present you as a confident, thoughtful expert in your field. Make sure to set a professional, concise, precise, and impactful tone. Because this is your work, I recommend using 'I' and being assertive. You can use sentences such as:

- 'I have been able to secure competitive grants', 'to build an international network', 'to explore novel ideas'.
- 'My key contributions are'; 'my technological innovations were recognized by…'
- 'I have leadership in the fields of…'; 'I am developing new expertise in…'

You can also show and demonstrate your own learning ability and how you improve, with statements such as:

- 'To improve my skills in…, I …'; 'I first noticed that …, then I …'

Finally, while you may want to develop a new technology, a new theory, a new product, train students and see them growing, I would suggest emphasizing subtly how your work contributes to the aim of your university. Universities are like businesses that want money, visibility, and to move up in the rankings. Therefore, any external grants or consultation that you get will give some funding back to the university, and will be favourably perceived. Similarly, publishing in top journals, being featured in the media, and getting awards will give the university a better visibility and an appealing public image, which will be viewed favourably. Finally, any outreach you may do that leads to more top students joining the university will also contribute to the ranking of the university and be favourable. While these are very pragmatic views, they can help you emphasize more how your research is impactful to the university and beyond, and will be helpful in positioning you as a potential tenured professor.

PERSONAL SHARING

I wrote the statements with the same kind of language I would use when writing a scientific paper in a high-impact journal and keeping in mind that the reviewers are professors, expert in their fields, who want to read something clear, concise, precise, and interesting. I tended to use 'I' although the research is done with the support of my research group.

KEY IMPORTANT POINTS

Set the tone:

- Write the statements to highlight and emphasize the importance and significance of your work.

10.3 Conclusions

Preparing the tenure dossier starts when the tenure clock starts and comprises the gathering of documents for all your activities, identifying referees, and writing statements that underline the impact and significance of your work. Understanding the expectations, timelines, and requirements in advance is necessary to optimize your time and maximize your impact. Examples of statements are provided in the additional references.

Additional References

Harvard office of faculty affairs: https://facultyresources.fas.harvard.edu/, Accessed 12/03/2021.

Guidelines for teaching portfolio, University of New Mexico: chrome-extension:https://digitalrepository.unm.edu/cgi/viewcontent.cgi?article=1197&context=assurance_argument_2018_19, Accessed 8/12/2022.

Guidelines for writing research, teaching and service statements, University at Albany: https://www.albany.edu/provost/promotion-tenure/apply/preparing-file/contents-file/statements, Accessed 30/11/2022.

Preparing yourself for the tenure process: 22 Feb 2010, https://serc.carleton.edu/NAGTWorkshops/earlycareer/tenure/preparing.html, Accessed 12/03/2021.

Writing a research statement for a tenure package: 08 July 2013, https://blogs.cornell.edu/danco/2013/07/08/writing-a-research-statement-for-a-tenure-package/, Accessed 17/10/2021.

J. Austin, Interdisciplinarity and tenure (2003). https://www.science.org/content/article/interdisciplinarity-and-tenure, Accessed 18/10/2018.

S. Pfirman, P. Balsam, R. Bell, P. Culligan, Maximizing productivity and recognition, Part 1: Publication, citation, and impact (2007). https://www.science.org/content/article/maximizing-productivity-and-recognition-part-1-publication-citation-and-impact, Accessed 18/10/2018.

S. Pfirman, R. Bell, P. Culligan, P. Balsam, J. Laird, Maximizing productivity and recognition, Part 2: Collaboration and networking (2008a). https://www.science.org/content/article/maximizing-productivity-and-recognition-part-2-collaboration-and-networking, Accessed 18/10/2018.

S. Pfirman, R. Bell, P. Culligan, P. Balsam, J. Laird, Maximizing productivity and recognition, Part 3: developing a research plan (2008b). https://www.science.org/content/article/maximizing-productivity-and-recognition-part-1-publication-citation-and-impact, Accessed 18/10/2018.

C.A. Trower, Perspective: advice on achieving tenure (2009). https://www.science.org/content/article/perspective-advice-achieving-tenure, Accessed 18/10/2018.

V. McGovern, Perspective: how to succeed in big science and still get tenure (2009). https://www.science.org/content/article/perspective-how-succeed-big-science-and-still-get-tenure, Accessed 18/10/2018.

11

Reflections on the Process and Outcome

11.1 Challenges and Hurdles during the Process and Paths to Overcome Them

The tenure-track process is a demanding and highly competitive journey requiring excellence in research, teaching, and service. Challenges and hurdles will come along the way, and finding mentors, getting feedback will be approaches to overcome them.

11.1.1 Finding Mentors

Joining a new university and meeting new people demand a little bit of time to adapt and build your network again. Finding trustworthy people who have the time and desire to mentor you and share their bits of knowledge can be a challenge. This is particularly true if you feel that you are part of a minority. Most universities and departments have a mentoring system, although to what extent it is effective is not clear.

To maximize your chances of getting hidden information, feedback, and helpful advice, it is a better plan to diversify your sources of information. You may reach out to some senior faculty to discuss your research and get acquainted. You can also befriend junior faculty who are a few years ahead of you. You can also tap into other formal mentoring or coaching services offered either within your university or externally. Finally, you can find numerous resources online, on other universities' websites, on personal blogs, or even watch career workshops on Youtube.

There is a large number of resources, and a few key ones are mentioned in this book in the additional references. While these resources may not be tailored specifically for your context, they provide plenty of information on what is behind the scenes. Besides, international universities competing for top ranking tend to have similar ways of functioning.

DOI: 10.1201/9781003635116-11

PERSONAL SHARING

I find it easier to get information through colleagues who are one step ahead because it is still very fresh in their minds and they have updated information as rules and expectations evolve with time. Having mentors can help with getting encouragement and not forgetting the long-term vision. However, for me, I find it difficult to name one person as my mentor. I would rather say that I have found a set of trusted people I can turn to, to ask for tips and advice. These trusted people may have been my past supervisors, or colleagues met through collaborations and committees, or even by casually chatting at conferences.

KEY IMPORTANT POINTS

Finding mentors:

- Informal information about the tenure process can be obtained through mentors and other support systems.

11.1.2 Getting Feedback

It is in general very difficult to get feedback from colleagues, except congratulations on achievements. Feedback may be welcomed for grants, the tenure dossier, or simply to know if you are on the right track. Feedback is difficult to get because (1) everybody is so busy giving feedback to students, doing peer review, etc. and (2) we start to be expert in our own specific domains and may not feel qualified to give feedback on someone else's work.

Feedback on your own overall trajectory can be received annually at the time of your performance review, as for any other staff. It may be very quick feedback, but can be right on the spot, objective, and very specific. Feedback on grant applications can be received from co-PIs or collaborators when writing the grant. Additional feedback can be received from the reviewers as well as from the grantor, upon request.

Feedback prior to journal paper submission may also be difficult to get, unless you add the name of the person who gave feedback to your authors' list. Feedback from the reviewers on journal papers can vary a lot, although it is generally given in a respectful manner.

Overall, you should find subtle ways to get indirect feedback through conversations. One way to identify if you are doing well and are on the right track can be (1) your gut feeling, (2) comparing yourself to others at the same career stage.

PERSONAL SHARING

I asked for feedback only on my teaching statement. I asked the person who led a workshop on the basics of teaching for faculty at the university. The positive feedback I received gave me confidence for my other statements. I like giving talks at conferences because this is when I can get informal feedback from the community on the quality, depth, and impact of the research. Some people also randomly give feedback, positive or negative. For example, you may receive a compliment (or a criticism) from a prominent colleague on a paper you published and posted on social media.

KEY IMPORTANT POINTS

Getting feedback:

- Seek feedback on your grant proposals, statements, overall CV and career achievements.
- Yearly performance reviews may be a great time to get feedback.

11.2 What Happens After You Get Tenure?

Getting tenure is one big step, but what is there after tenure? This section discusses about receiving the news and how your position might change after.

A few months after submitting your tenure dossier, you should receive the news. If the news is not a positive one, you will be provided with more specific details and you will have a chance to re-apply in the future. Make sure you understand what was missing and have a discussion with your department Chair to plan for the successful re-application. Due to the tenure clock, you may be granted only one second trial.

If it is successful, congratulations! Some departments have a culture of celebration, with an aperitif or dinner. Make sure to celebrate with your team, students, friends, family, etc. You have achieved a big step. You can expect to receive a new contract to crystallize your tenure and your position in the university. The job security that comes with tenure can be a relief, as it ensures job stability. After the announcement, an official letter will be issued along with a new contract. While many will congratulate you, it is not guaranteed you will receive any specific feedback.

What happens next after tenure and promotion to associate professor? The next and final stage is to obtain full professorship, which has its own set of requirements. The transition to full professor is less clear than the tenure process and mentorship and guidance are a lot more challenging to find. If you want to go down that path, you can start again looking for information with senior faculty, Chair, etc.

Also, signing a new contract of many years comes with questions on what your goals are and the need to re-think your career and aims. Likely your role will also change with the promotion and getting higher in the hierarchy. More administrative work will come your way and may shift your research dynamics.

PERSONAL SHARING

I received the successful announcement from our school Chair a few hours before it was announced in the school meeting. The next day, I was asked to join another committee and do more admin work. I was not expecting to have this change of focus. As an assistant professor, I felt very protected from other commitments and extra demands. Now, as an associate professor, this protection is no longer there, and it becomes even more challenging to balance research, teaching, and service. In addition, while I was of course very happy to get the tenure, thinking of the next 30 years gives a very different perspective: what do I want to do in the next 30 years? Where do I want to be? How and where do I see myself in 10, 20, or 30 years? I was also surprised to learn that the grant application landscape becomes more challenging, with the expectation for bigger grants and more collaborative grants.

KEY IMPORTANT POINTS

What happens after you get tenure:

- You will receive the announcement with or without feedback.
- There is always the opportunity to ask for more details and to re-apply if necessary.

11.3 Learnings from the Tenure-Track Process

The tenure-track lasts about 5–7 years, and has ups and down. However, it is also a learning process. Let's review here what are the key learnings from the tenure-track.

One first learning is to get organized, to set priorities, and to juggle tasks efficiently. You learn how to develop a storyline for your research that goes beyond a single short-term research project. You also learn how to define a long-term (5 year-ish) research vision and goals to achieve this vision.

You also learn leadership skills, which you practice at least for your own research group. You show you can encourage and motivate others along your research goals. You learn how to articulate and defend your ideas.

You learn how to teach effectively and how to mentor students and researchers, in the classroom and in the laboratory. You become an educator.

You develop a network, which allows you to have a global vision of the current research trends and gaps, which enables you to further define your

PERSONAL SHARING

Thanks to the tenure track process, I feel my most important learning is about seeing the overall direction where research is going, internationally as well as locally. I also feel more able to tell that certain places (countries or universities) would be better to do certain types of research. I feel also more able to say 'there is little research in this area and this would be promising'. I feel that being in academia in a research-intensive international university with global connections allows me to have this overview and imagine ahead of time what we could do as a research community to address specific challenges. I am also now very clearly aware of my own limitations, for example that there is a broad gap between my research and industrial applications and that collaborating with industry may not be for me. I also discovered how I enjoyed enabling people and opening doors, but not necessarily taking people by the hand. Overall, I am quite impressed by how much we can accomplish during the tenure-track.

KEY IMPORTANT POINTS

Learnings from the tenure-track process:

- So many learning opportunities on so many levels.

own research direction. You are also able to advise others and provide them with constructive feedback.

You also discover what you like and what you do not like in the academic world.

11.4 Conclusions

Having a support network with mentors, colleagues, friends able to give advice and feedback is necessary to overcome challenges and hurdles along the tenure track. After you get tenure, your perspective on your own research, teaching and service may change as well as your role in these areas. Overall, the tenure-track is not just a competitive and challenging way to get a permanent job, it is also a learning process that enables you to become a world leading researcher, educator, and to learn more about yourself.

Additional References

L. Catanese, A. Shoamanesh, Identifying the right mentor, Stroke 48 (2017), e248–e251.

S. Sarabipour, S.J. Hainer, F.N. Arslan, C.M. de Winde, E. Furlong, N. Bielczyk, N.M. Jadavji, A.P. Shah, S. Davla, Building and sustaining mentor interactions as a mentee, FEBS J 289 (2022), 1374–1384.

C. Ocobock, A. Niclou, T. Loewen, K. Arslanian, R. Gibson, C. Valeggia, Demystifying mentorship: tips for successfully navigating the mentor-mentee journey, Am J Hum Biol 24 (2022), e23690.

12

Conclusion

12.1 Concluding Words

The tenure-track is a learning journey to independence and one step towards becoming a full professor. While the focus is excellence in research, teaching, and service, many opportunities come to develop specific areas of interest and to build one's own community. In this book, we have seen the diversity of tasks that an assistant professor has to conduct efficiently. We have discussed about establishing and leading a research group, supervising students and researchers, getting funds, teaching, writing papers and getting published, while maintaining a work-life balance and ensuring diversity. We have also looked at the details of the tenure-track dossier and what comes after the tenure.

I hope the content of this book is able to provide you with a more concrete idea about the tenure-track in STEM fields and that the additional references can be of some help to get more information. It should be emphasized again that each institution is different and has its specific expectations. Everybody also has their own experience. At the same time, we also see new general current trends in academia to depart from the quantitative measurements such as h-index, impact factors, etc. towards a more qualitative approach of excellence, which includes awards, leadership, etc. While it is important to understand the expectations and fulfil them, it is also extremely important to find inner motivation and goals. Not everybody wants to be an academic and even after tenure, some may still change careers to fulfil other goals. This book provides my own views and experience and I urge you to seek other views to complement them.

Finally, the tenure track is still today one prestigious academic route. The reviewing by established peers, the excellence, and the aims of the tenure are acknowledged. Nevertheless, the tenure process may still have some pitfalls and biases. It is not sure if it is here to stay forever or will evolve. Some universities want to revoke this process while some other countries do not have it at all and instead use multiple promotion steps. There are therefore many alternatives to being in academia and conducting research, teaching and service so that anyone can find their place. In addition, the new technologies

DOI: 10.1201/9781003635116-12

such as artificial intelligence (AI) are already disrupting how people connect, learn, and work. In academia, AI may facilitate the writing of papers and grants, as well as allowing more accessible knowledge. We now see new types of autonomous and self-driven laboratories. As a future researcher or professor, how do we ensure we are needed and unique? Multidisciplinarity, creativity, and analysing skills are likely to be even more important to master to stay ahead of this accelerated research and development.

I hope you enjoyed the book and that you will be successful in what you set yourself to do. Please check out the additional references to get more specific tips or other views on the topics discussed in the book.

Additional References

M. Pietila, R. Pinheiro, Reaching for different ends through tenure track – institutional logics in university career systems, Higher Educ 81 (2021), 1197–1213.

A. Ashcraft, J.S. Andersen, M.M. Rogge, H. Song, L. Opton, Academic tenure: perceptual variations among tenured, tenure-seeking and non-tenured faculty, J Prof Nursing 37 (2021), 578–587.

J.B. Main, Y. Wang, L. Tan, The career outlook of engineering PhDs: influence of postdoctoral research positions on early career salaries and the attainment of tenure-track faculty positions, J Eng Educ 110 (2021), 977–1002.

T.P. Clement, How can we reform the STEM tenure system for the 21st century, PNAS 119 (2022), e2207089119.

D. Dawson, E. Morales, E.C. McKiernan, L.A. Schimanski, M.T. Niles, J.P. Alperin, The role of collegiality in academic review, promotion, and tenure, PLoS One 17 (2022), e0265506.

N.A. Bonn, W. Pinxten, Advancing science or advancing careers? Researchers' opinions on success indicators, PLoS One 16 (2021), e0243664.

Index

Note: **Bold** page numbers refer to tables, *italic* page numbers refer to figures.

ENERGY POLICIES, POLITICS AND PRICES

Additional books in this series can be found on Nova's website under the Series tab.

Additional e-books in this series can be found on Nova's website under the e-book tab.

ENERGY POLICIES, POLITICS AND PRICES

MEXICO'S OIL AND GAS INDUSTRY

CONSIDERATIONS, REFORM ISSUES, AND U.S. INTERESTS

ENERGY POLICIES, POLITICS AND PRICES

MEXICO'S OIL AND GAS INDUSTRY

CONSIDERATIONS, REFORM ISSUES, AND U.S. INTERESTS

NEIL BECKER
EDITOR

nova publishers
New York

Library of Congress Cataloging-in-Publication Data

ISBN: 978-1-63483-846-7

Published by Nova Science Publishers, Inc. † New York

CONTENTS

PREFACE

The future of oil and natural gas production in Mexico is of importance for both Mexico's economic growth, as well as for U.S. energy security, a key congressional interest. Mexico is a top trade partner and the 3rd largest crude oil supplier to the United States. Mexico's state oil company, Petroleos Mexicanos (Pemex) remains an important source of government revenue even as it is struggling to counter declining oil production and reserves. Due to an inability to meet rising demand, Mexico has also significantly increased natural gas imports from the United States. Still, gas shortages have hindered the country's economic performance. This book provides an overview of Pemex and the content and prospects for Mexico's energy reforms, before discussing specific issues facing Mexico's oil and gas industry. It then examines the U.S.-Mexico energy relationship through the lenses of trade and energy cooperation. It concludes by suggesting several oversight issues for Congress related to what the enactment of energy reform might portend for Mexico's economic development, the U.S. energy matrix, and bilateral or North American energy cooperation.

In: Mexico's Oil and Gas Industry
Editor: Neil Becker

ISBN: 978-1-63483-846-7

Chapter 1

MEXICO'S OIL AND GAS SECTOR: BACKGROUND, REFORM EFFORTS, AND IMPLICATIONS FOR THE UNITED STATES*

Clare Ribando Seelke, Michael Ratner, M. Angeles Villarreal and Phillip Brown

SUMMARY

The future of oil and natural gas production in Mexico is of importance for both Mexico's economic growth, as well as for U.S. energy security, a key congressional interest. Mexico is a top trade partner and the 3rd largest crude oil supplier to the United States. Mexico's state oil company, *Petroleos Mexicanos* (Pemex) remains an important source of government revenue even as it is struggling to counter declining oil production and reserves. Due to an inability to meet rising demand, Mexico has also significantly increased natural gas imports from the United States. Still, gas shortages have hindered the country's economic performance.

On December 20, 2013, Mexican President Enrique Peña Nieto signed historic constitutional reforms related to Mexico's energy sector

* This is an edited, reformatted and augmented version of a Congressional Research Service publication R43313, prepared for Members and Committees of Congress, dated July 30, 2015.

aimed at reversing oil and gas production declines. On August 11, 2014, secondary laws to implement those reforms officially opened Mexico's oil, natural gas, and power sectors to private investment. As a result, Pemex can now partner with international companies that have the experience and capital required for exploring Mexico's vast deep water and shale resources. Leftist parties and others remain opposed to the reforms.

The energy reforms transform Pemex into a "productive state enterprise" with more autonomy and a lower tax burden than before, but make it subject to competition with private investors. They create different types of contracts for private companies interested in investing in Mexico, including production-sharing and licensing; allow companies to post reserves for accounting purposes; establish a sovereign wealth fund; and create new regulators. In December 2014, the Mexican government announced the terms of part one of Round 1, under which shallow-water offshore exploratory blocks available for public bidding would be auctioned. On July 15, 2015, Mexico's Energy Ministry announced that only 2 of the 14 blocks available were awarded to successful bidders. The government is likely to alter the terms offered, including the amount of investment required by companies, in order to attract more interest in subsequent bidding rounds.

The U.S. Congress has legislative and oversight interests in examining the implications of Mexico's oil and natural gas reforms on U.S. hydrocarbons imports and exports, bilateral trade and investment, and economic conditions in Mexico. In December 2013, Congress approved the U.S.-Mexico Transboundary Hydrocarbons Agreement (P.L. 113-67), which aims to facilitate joint development of oil and natural gas in part of the Gulf of Mexico. The 114th Congress is likely to consider legislative proposals to speed up energy infrastructure development, including cross-border natural gas pipelines, as was approved by the House on January 21, 2015 (H.R. 161). Congress may also consider proposals to revise U.S. crude oil export policy.

The opening of Mexico's oil and natural gas sector could expand U.S.-Mexico energy trade and provide opportunities for U.S. companies involved in the hydrocarbons sector, as well as infrastructure and other oil field services. If these reforms accelerate growth and investment in Mexico (as the government has stated), they could also benefit North American competitiveness. Industry analysts maintain that the reforms are generally well-designed, but that the way they are implemented will likely determine their impact. The success of the reforms may also depend on trends in global oil prices. Should oil prices remain at current levels, shale resources and other unconventional fields may not be feasible to develop at this time.

INTRODUCTION

The United States has a strong economic interest in ensuring energy security, bolstering exports, and reducing barriers to U.S. trade and investment. The United States also has a national security and an economic interest in ensuring that Mexico, a key ally and top trade partner with which the United States shares a nearly 2,000 mile border, is economically vibrant and politically stable.[1] U.S.-Mexico energy trade and cooperation plays an important role in achieving those goals. Energy cooperation has become a priority of bilateral economic cooperation under the U.S.- Mexico High Level Economic Dialogue as well as a focus of trilateral research, planning, and coordination among the energy secretaries of the United States, Canada, and Mexico.[2]

U.S. policymakers are closely monitoring the implementation of the December 2013 constitutional reforms and August 2014 implementing laws that, among other things, allow Mexico's *Petroleos Mexicanos* (Pemex) to partner with international companies and other industry entrants to boost Mexican production. Hailed by many analysts as the most significant economic reform undertaken by Mexico since its entrance into the North American Free Trade Agreement (NAFTA) in 1994, the energy reforms are expected to boost investment, growth, and eventually oil and gas production in the country. The reforms also opened Mexico's electricity sector to private generators, although not addressed in this report.[3] If power sector reforms can help reduce Mexico's high electricity costs, then Mexico's manufacturing sector—a dynamic sector that is highly integrated with U.S. industry—should become more competitive.[4]

There is a danger, however, that the Mexican government may have oversold the benefits of the reforms. Investors initially appeared to have maintained interest in Mexico's energy sector despite declines in oil prices, but the results of Mexico's first round of public bidding announced in July 2015 proved disappointing.[5] At a House hearing held on July 23, 2015, witnesses cited the limited size and production potential of the blocks offered, the high minimum bid required, and investors' concerns about corruption and insecurity as possible reasons for the lack of interest in the tender.[6]

This report provides an overview of Pemex and the content and prospects for Mexico's energy reforms, before discussing specific issues facing Mexico's oil and gas industry. It then examines the U.S.-Mexico energy relationship through the lenses of trade and energy cooperation. It concludes by suggesting several oversight issues for Congress related to what the enactment of energy reform might portend for Mexico's economic development, the U.S. energy matrix, and bilateral or North American energy cooperation.

PEMEX: A BRIEF HISTORY AND PRE-REFORM STATUS REPORT

Foreign investment in Mexico's oil industry has had a tumultuous history.[7] After oil was discovered in Mexico at the turn of the 20th century, foreign investors—primarily from Britain and the United States—played a significant role in helping the country become the world's second largest oil producer by the early 1920s. However, political unrest during and after Mexico's bloody revolution (1910-1920) and the country's 1917 constitution, which established national ownership of all hydrocarbons resources, caused investment in Mexico's oil and natural gas sectors to gradually decline. By the 1930s, reduced foreign investment had resulted in dramatic declines in production levels, and fraught relations between U.S. oil companies and successive post-revolutionary presidents had damaged U.S.-Mexican relations. Tensions culminated in President Lázaro Cárdenas' historic 1938 decision to abandon efforts to mediate a bitter labor dispute between Mexican oil workers and foreign companies and instead follow through on his threat to expropriate all U.S. and other foreign oil assets in Mexico.

Upon its creation in 1938, Pemex became a symbol of national pride and a rallying point around which Cárdenas and what became the Institutional Revolutionary Party (PRI)—the party of President Enrique Peña Nieto—united a disparate Mexican society against foreign intervention. Oil remains deeply tied to Mexican nationalism. Nevertheless, Pemex continued to pursue service contracts with some U.S. oil companies until the practice was definitively outlawed by a 1958 regulatory law implementing Article 27 of the constitution.[8] From then on, Pemex retained a monopoly over Mexico's oil and natural gas sector and the Mexican Finance Ministry kept tight control over the companies' finances and management.

In 2013, seventy-five years after its founding, Pemex found itself facing significant challenges. Pemex had its heyday in the late 1970s following the discovery of the huge shallow water Cantarell oil field, but the company's long-term performance had been hindered by a number of factors. For years, Pemex sustained itself on the revenue produced from its relatively easy-toexploit shallow water fields without investing the capital necessary to replace those reserves with new fields or even maintain its infrastructure. Pemex had a high percentage of losses, low worker productivity, and facilities that are in significant need of repair; 37 people were killed in January 2013 after an explosion occurred at one of the company's offices in Mexico City.[9] In part because of the Mexican government's heavy tax demands, Pemex had operated at a loss since 1998 and significantly increased its debt burden. Until recently, the government had also prevented the company from reinvesting its profits into maintenance and new exploration.[10] Pemex's pension liabilities, negotiated by the company's powerful and, for some observers, corrupt workers union, had become an unsustainable drain on its finances.[11] Each year, Pemex had been losing hundreds of millions of dollars due to criminal groups illegally tapping into its pipelines.[12] Pemex's inability to partner with other companies arguably inhibited it from benefitting from new expertise and techniques, particularly in deep water drilling and hydraulic fracturing (fracking).

THE PATH TO PEMEX REFORM

Proposals for Energy Reform

President Enrique Peña Nieto of the nationalistic PRI assumed the Mexican presidency on December 1, 2012 after 12 years of rule by the conservative National Action Party (PAN). Even though Peña Nieto hailed from the PRI, the party that had nationalized the oil industry and watered down previous PAN-led reform efforts, he campaigned on an economic platform that prioritized allowing Pemex to form joint ventures with private companies. At his inauguration, President Peña Nieto announced a reformist agenda aimed at bolstering Mexico's competitiveness that included energy sector reform.

In order to implement his agenda, President Peña Nieto and leaders of the PRI concluded a "Pact for Mexico" agreement with the PAN and the leftist Party of the Democratic Revolution (PRD) that facilitated the passage of financial, education, telecommunications, and fiscal reforms.[13] Many of those reforms required constitutional changes. As with the other constitutional reforms enacted in 2013, the energy reforms that Peña Nieto proposed in August of that year required a two-thirds vote in the Mexican Congress and approval by a majority of the country's 32 state legislatures. As discussed below, a PRI-PAN alliance enabled the December 2013 approval of constitutional reforms on energy, but led the PRD to leave the Pact for Mexico.

The Peña Nieto Administration's August 2013 energy reform proposal would have removed hydrocarbons from the list of strategic sectors that can only be developed by the government and allowed Pemex to form "profit-sharing"[14] partnerships with international companies in exploration and production. The reform would also have allowed Pemex to sign agreements with private companies for transporting oil and gas, refining, and producing petrochemicals. In secondary legislation, the President pledged to introduce reforms to give the company budget autonomy, improve its transparency, and change its fiscal structure, among other measures.

The PAN put forward deeper reforms than the PRI that would permit private concessions in upstream and downstream[15] operations and production-sharing agreements between Pemex and private companies. Former PAN President Felipe Calderón tried to enact far-reaching energy reforms in 2008, but his proposal was watered down by the PRI-led Congress.[16] The PAN also sought to establish a strong regulatory body and a sovereign wealth fund to support social needs. Some PAN legislators reportedly conditioned their support for energy reforms on the PRI backing political reforms, which were also approved in December 2013.

The PRD vigorously opposed allowing private involvement in Pemex. Instead the PRD proposal focused on reforming the company while simultaneously granting it greater budget autonomy and a less onerous tax burden. Past and present PRD party leaders joined forces to oppose the PRI and PAN versions of energy reform.

Key Provisions of the December 2013 Constitutional Reforms[17]

In the end, the reforms approved by the Mexican Congress[18] and a majority of state legislatures and then signed into law by President Peña Nieto on December 20, 2013, bore most in common with the PAN proposal, but contain elements of all three parties' energy reform propositions.

Key elements of the reforms include:

- Maintaining state ownership of subsoil hydrocarbons resources, but allowing companies to take ownership of those resources once they are extracted and to book reserves for accounting purposes;

- Creating four types of contracts for exploration and production: service contracts (companies are paid for activities done on behalf of the state), profit-sharing contracts, production sharing contracts, and licenses (enabling a company to obtain ownership of the oil or gas at the wellhead after it has paid taxes);
- Opening refining, transport, storage, natural gas processing, and petrochemicals sectors to private investment;
- Transforming Pemex into a productive state enterprise with an autonomous budget and a board of directors that does not include union representatives;
- Strengthening four federal entities with regulatory roles in the hydrocarbons industry (the Ministries of Energy and Finance, the National Hydrocarbons Commission or CNH, and the Energy Regulatory Commission) and creating a National Agency for Industrial Safety and Environmental Protection; and
- Establishing a sovereign wealth fund, the Mexican Petroleum Fund for Stabilization and Development, to be managed by the Central Bank. (See *Figure 1* below.)

Secondary Laws Approved in August 2014[19]

While the constitutional reforms outlined the broad contours of Mexico's oil and gas reform, many details were left to be defined in secondary laws needed to implement those reforms. Those laws had to pass Congress by a simple majority and technically had to be enacted within 120 days of the signing of the energy reforms (December 20, 2013). However, due to a variety of factors, including the PAN's internal elections held in May 2014, the Mexican Congress did not consider the energy laws until a special section convened in July. The Senate and Chamber of Deputies debated and made minor modifications to the Peña Nieto Administration's original proposals that were then approved with broad support from the PRI, PAN, and smaller parties. Key provisions dealing with the hydrocarbons sector not mentioned above include:

- Pemex: is more independent of the state, but must adopt internal reforms; is to pay a tax rate of roughly 65% (rather than 79% now);

and is permitted to keep some of its existing fields through a "Round Zero" process as deemed appropriate by the Secretariat of Energy;

- Pemex's monopoly on retail gasoline and diesel sales ends in 2016;
- Companies will pay royalties and taxes varying with the price of oil;
- Companies must respect national content requirements of 25% in 2015, rising to 35% in 2025 (excluding deep water activities);
- Companies may not expropriate land from communities for exploration and development, but rather temporarily occupy land and compensate its owners; and,
- The National Hydrocarbons Commission (CNH) is strengthened and established as a constitutionally coordinated entity to gather and manage information on the energy sector; issue regulations and monitor compliance; and manage the bidding rounds, award contracts, and supervise those contracts.

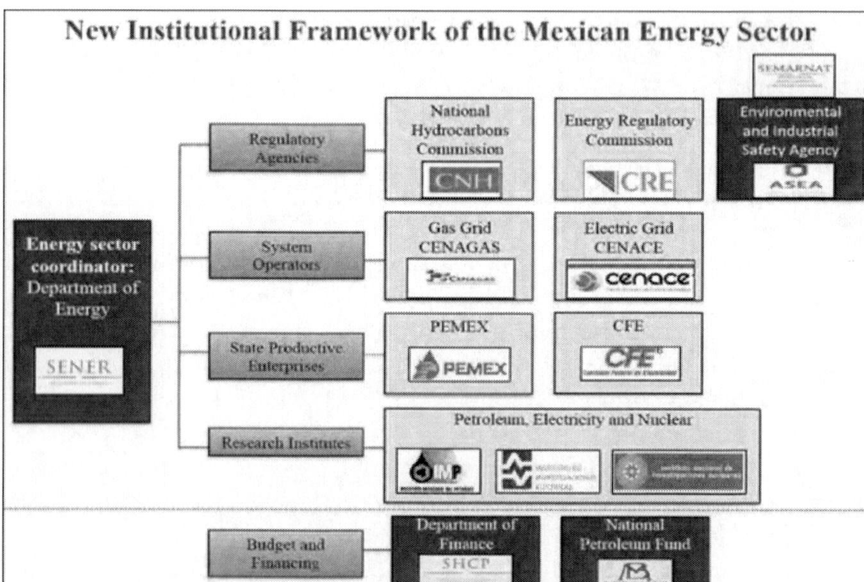

Source: Government of Mexico.

Figure 1. Mexico's Hydrocarbons Sector.

The institutions now in charge of Mexico's hydrocarbon industry[20] now include:

- *Ministry of Energy (SENER):* develops Mexico's upstream policy; determines areas to be made available and the schedule for public bidding; chooses which of the contract models to apply to which contract; and approves the non-fiscal terms of the contract.
- *Ministry of Finance (SHCP):* determines the fiscal terms to apply to each contract and participates in audits.
- *National Hydrocarbons Commission (CNH):* interfaces with Pemex and private companies, conducts and manages contracts, and oversees the industry.
- *Energy Regulatory Commission (CRE):* grant permits for transportation, storage, distribution, compression, liquefaction, decompression, regasification, marketing, and sale of crude oil, oil products, and natural gas.
- *National Agency for Industrial Safety and Environmental Protection (new):* regulates environmental and safety concerns.
- *National Natural Gas Control Center (CENAGAS):* manages system for gas distribution and storage.

Round Zero

On August 13, 2014, the Mexican government announced the results of Round Zero a month earlier than had been required by the legislation. The Secretariat of Energy awarded Pemex 83% of Mexico's probable reserves and 21% of its prospective reserves. Pemex had requested to retain 31% of the country's prospective reserves, but its limited capacity to explore and produce in deep water and unconventional areas may be why it was not awarded that large a share.

Even though Round Zero did not open up new fields for private companies to bid on, it did increase the opportunities for private involvement in exploration and production in Mexico., Pemex is in the process of changing some existing service contracts it has with oil companies (as allowed under the 2008 reforms) into profit-sharing contracts, as well as seeking new partners to help it develop some of the other probable and prospective fields. New companies interested in investing in Mexico may first seek to partner with Pemex before submitting independent bids.

Ongoing Developments and Round One

In the past year, Pemex has been undergoing restructuring; all of the regulatory agencies empowered or created by the energy reforms have begun hiring and training their staffs; regulations have been released; and the priorities for the Mexican Petroleum Fund have been defined.[21] U.S. companies are closely watching the degree to which Mexico establishes sound environmental and safety standards, recruits qualified individuals to staff regulatory agencies, fosters transparency, and conducts fair and transparent bidding processes. These efforts have arguably been made more difficult in recent months due to the fact that Mexico's 2015 budget— including funding for those energy-related entities—has been cut by some $8 billion due to declining oil prices (and therefore declining government revenue).[22]

On December 14, 2014, the Mexican government announced the terms under which companies already incorporated in the country could bid on production-sharing contracts to explore 14 shallow-water fields of prospective light crude in the Gulf of Mexico. The Mexican government has maintained that the fields would be profitable even if oil prices were to fall below $50 per barrel; average production costs for Pemex hover around $22/barrel.[23] On July 15, 2015, Mexico's Energy Ministry announced the bidding results. The results have been deemed disappointing by energy analysts, as only 2 of the 14 blocks available were awarded to successful bidders.[24]

While analysts concur that there is still interest in Mexico's remaining "round one" offerings even though oil prices are low, they predict that the government will have to offer more competitive terms in order to garner greater interest. Aggressive drilling requirements in the proposed contracts may also deter companies in the current low price environment. The next rounds expected to be launched in the coming months include bidding for shallow water production and mature fields.

MEXICO'S ENERGY RESOURCES

Oil: Attracting a Lot of Interest

Mexico is the world's 10th largest producer of oil and holds approximately11.1 billion barrels of oil reserves—the 18th largest in the world.[25] Mexico may also have the 8th largest tight oil[26] resources globally, about another 13 billion barrels.[27] With these reserves, Mexico has the potential to halt its decade-long decline in oil production.

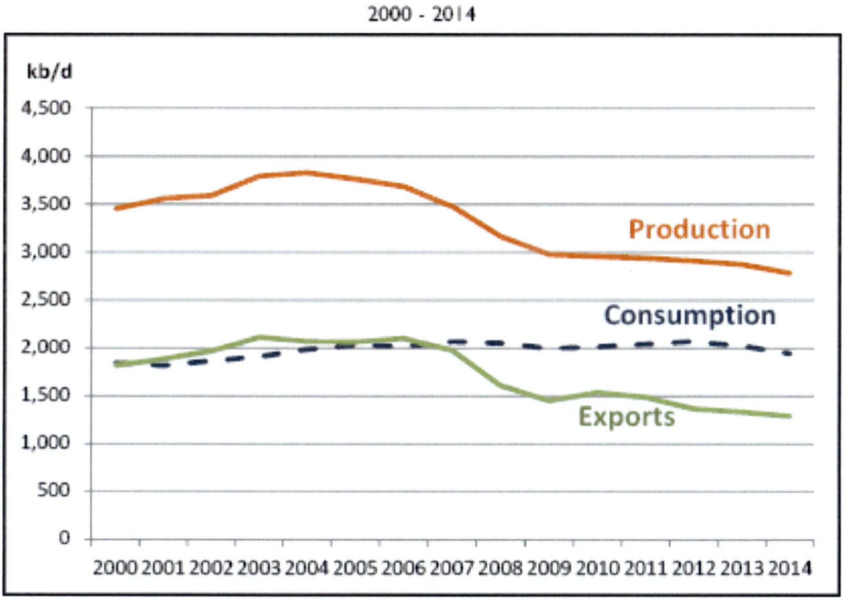

Source: BP Statistical Review of World Energy 2001 - 2015.
Notes: Units = thousand barrel per day (kb/d).

Figure 2. Mexican Oil Production, Consumption, and Exports.

Mexico's oil production declined by some 20% from 2005 through 2009; production has fallen by roughly 1% per year since that time (see *Figure 2*) due, in part, to aging and inefficient infrastructure (see *Figure 3*).[28]

Nevertheless, Mexico lags only behind Russia, the United States, China, and Canada as an important non-OPEC oil producer. Most of Mexico's production (75%) is found offshore in the shallow waters of the Bay of Campeche, which is part of the Gulf of Mexico, and concentrated in two fields—Ku-Maloob-Zaap (KMZ) and Cantarell. KMZ production has been on the rise since 2006, reaching almost 864,000 barrels per day (b/d) at the end of 2013, and has replaced part of Cantarell's decline. Cantarell was once one of the largest producing fields in the world, but started having pressure problems in the mid-1990s. Efforts to reverse the decreasing production were successful for a while and the field reached its peak in 2004 at 2.1 million b/d (63% of

Mexico's production). By 2013, Cantarell produced only about 440,000 b/d (17% of Mexico's production).

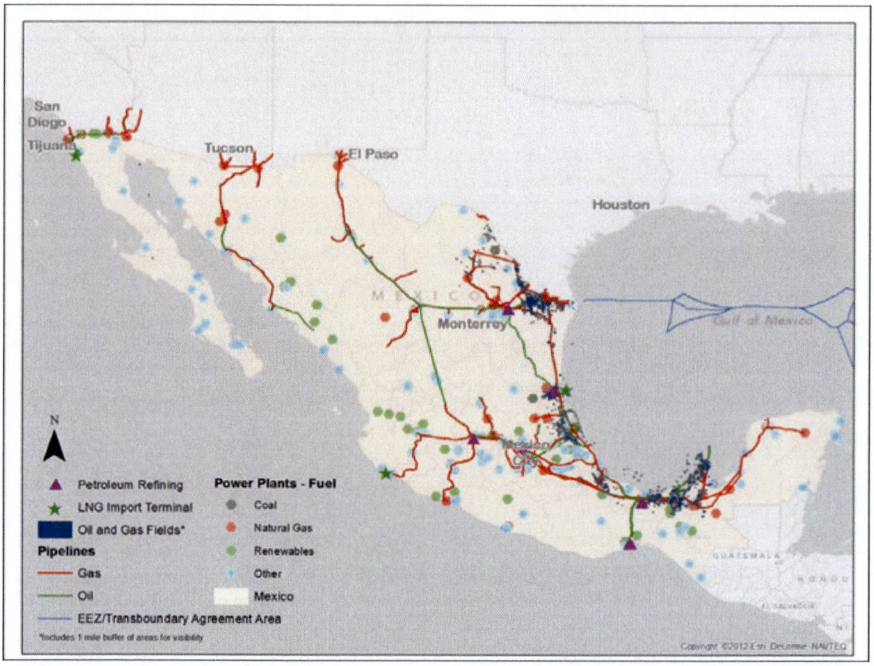

Source: Compiled by CRS using data from IHS, Platts, and Esri. Date: September 2013.

Figure 3. Mexico Energy Infrastructure.

The U.S. Energy Information Administration (EIA) has estimated that the recently-enacted energy reforms could boost Mexico's long-term oil production potential to 3.7 million b/d by 2040 from the 2.9 million b/d produced in 2013. That estimate is 75% higher than the EIA's 2013 forecast for Mexico's long-term oil production that was issued prior to the enactment of energy reforms.[29]

There may also be significant deep water resources in the Gulf of Mexico yet to be discovered. As can be seen below in *Figure 4*, Mexico has undertaken very little activity in its portion of the Gulf of Mexico, particularly compared to the United States, in part because Mexico does not yet have the technical capacity to effectively explore or produce its deep water areas. This is one of the reasons that international companies, particularly those with deep water expertise, are excited about the reforms in Mexico. Additionally, the U.S.-Mexico Transboundary Agreement (see below) may play an important role in raising Mexico's standards of operation in deep water.

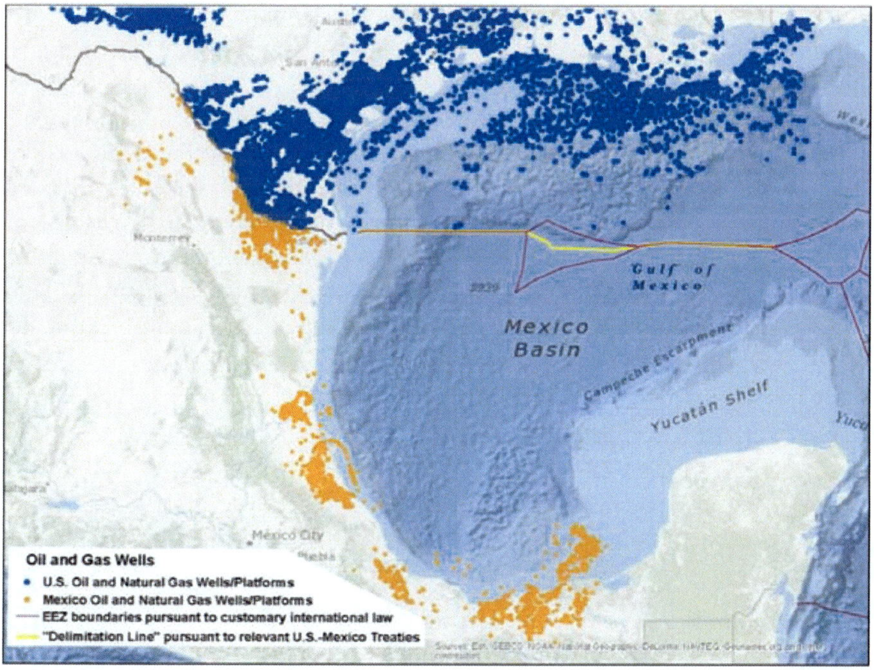

Source: Compiled by CRS using data from HSIP Gold 2012 (Platts), IHS 2012 Wells, and Esri.

Note: This map may or may not offer information related to the submerged land tenure systems of the United States or Mexico.

Figure 4. U.S./Mexico Oil and Natural Gas Activity Around the Gulf of Mexico. (Active Wells and Platforms)

Natural Gas: More Needed

Mexico's natural gas production capacity is higher than in 2000, but has not been able to keep up with demand (see *Figure 5*), which increased about 123% between 2000 and 2014. Most of Mexico's natural gas consumption supports its oil operations and national electricity generation. Mexico's proven gas reserves are on the decline due to underinvestment in exploration. Production has also begun to decline in recent years as price differentials have made it more profitable for Pemex to produce oil than gas. And, although Mexico may have significant unconventional natural gas resources, it is further behind in developing these resources than other countries, such as Canada. Mexico's energy reforms seek to attract private investment in exploration and production of both conventional and unconventional natural gas resources, but also to accelerate investment in much-needed infrastructure for storage, transportation, and distribution.

As a consequence of demand rising faster than production, Mexico's imports of natural gas have also been increasing, accounting for about 35% of consumption today compared to less than 10% in 2000. In 2000, Mexico imported 30% of U.S. natural gas exports which accounted for 100% of Mexico's natural gas imports. In 2006, Mexico started importing liquefied natural gas (LNG) at very high costs from Qatar, Nigeria, and Peru to help meet its growing demand for gas. Mexico has three LNG import terminals, two on the Pacific side and one on the Atlantic.

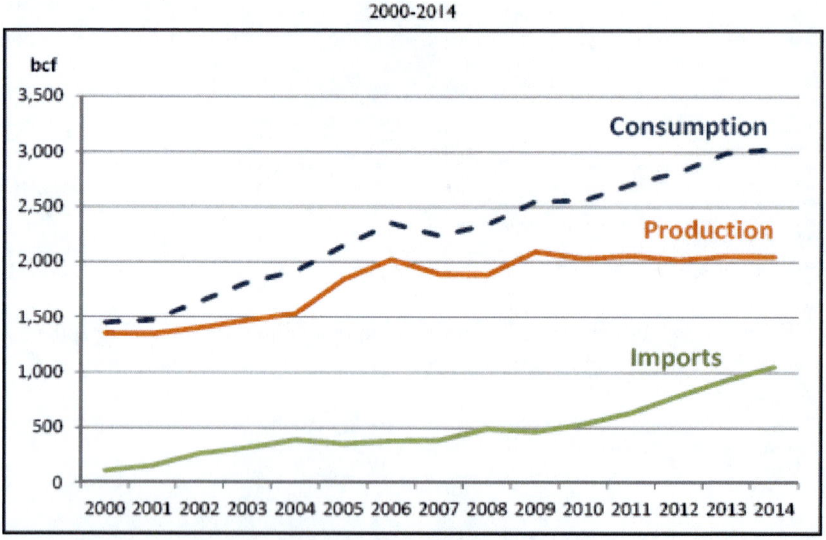

Source: BP Statistical Review of World Energy 2001 - 2014.
Notes: Units = billion cubic feet (bcf).

Figure 5. Mexican Natural Gas Production, Consumption, and Imports.

As a free trade partner, exports of U.S. natural gas to Mexico are assumed in the public interest by U.S. statute and permitted without delay, which has spurred U.S. natural gas exports to Mexico. In 2014, U.S. natural gas exports accounted for approximately 69% of Mexico natural gas imports, and 24% of its natural gas consumption. Between 2008 and 2013, U.S. pipelines to Mexico doubled their capacity. Current U.S. export capacity to Mexico is 4.9 billion cubic feet (bcf) per day. By the end of 2015, that capacity should reach 8 bcf/d.[30] It would appear that for the immediate future Mexico will likely remain dependent on U.S. supplies of natural gas to meet its growing demand.

Unconventional Oil and Natural Gas Opportunities

One of the areas gathering interest with the opening of Mexico's oil and natural gas industry has been shale development. The EIA has assessed Mexico's shale gas resources to be significant (the 6th highest globally).[31] The proximity of some formations in northern Mexico to U.S. developments makes them attractive to some U.S. companies. As an example, the Eagle Ford basin in Texas, one of the fastest growing shale producing areas in the United States, may extend down into Mexico.

Mexico, through Pemex, has already started exploring some of its unconventional formations. A limited number of test wells have been drilled, but Pemex has ambitious plans for scaling up development and production over the next 10 years. However, Mexico will need to implement reforms to attract outside investment to strengthen regulatory and environmental protection measures; expand pipelines, roads, and other infrastructure; address water management issues; and deal with security concerns. Some of the states in northeastern Mexico where shale formations are located have experienced significant violence in recent years, possibly deterring additional business opportunities. The impact of government efforts to quell that violence and criminality in those states has been limited.[32] Potential disputes with individuals and communities over access to land could for exploration and development, as well as opposition to fracking in some areas, could also deter would-be investors. The United States has already been working with Mexico in some of the more technical areas such as resource assessment, environmental protection, and regulatory policies.

Refining: Limited Capacity and in Need of Modernization

Although Mexico is a large exporter of crude oil, it is a net importer of refined petroleum products, such as gasoline and diesel fuel. Mexico does not have enough refining capacity of its own to meet its domestic demand for refined products, nor has it made the investment to process heavy crudes like its Maya crude. Mexico has six refineries with a total capacity of 1.54 million barrels per day, but in recent years has operated below capacity because of operating mishaps.[33] As in many other countries throughout Latin America, Mexico's refineries are in need of major repairs and upgrades and often operate at below their stated capacity.[34] Mishaps and other losses are expected to result in a $7.7 billion loss for the company in 2013.[35]

Mexico and the United States already have a close relationship in the refining sector. Much of the U.S. Gulf Coast refining capacity is designed to process heavy crudes, which require more sophisticated and expensive technologies than Mexican refineries currently possess. Mexico exports its heavy crude to U.S. refineries on the Gulf Coast, which then sends some of the refined products back to Mexico. Pemex, which operates all of the refineries in Mexico, also owns 50% of a refinery in Texas. The refining relationship between Mexico and the United States could potentially be expanded even further as the reforms fully open up Mexico's downstream (marketing and refining) hydrocarbons market to international companies.

ENERGY: A CENTRAL COMPONENT OF U.S.-MEXICO TRADE

The bilateral economic relationship with Mexico is of key interest to the United States because of Mexico's proximity, the high volume of trade with Mexico, and the strong economic ties between the two countries. The United States is, by far, Mexico's leading partner in merchandise trade, while Mexico is the United States' third largest trade partner in total trade after China and Canada. Mexico is the United States' second largest export market after Canada and ranks third as a supplier of U.S. imports.

Since NAFTA took effect in 1994, the United States and Mexico have become more economically integrated with strong trade and supply chain linkages. U.S. exports to Mexico increased rapidly since NAFTA, increasing from $41.6 billion in 1993 to $240.3 billion in 2014, an increase of 478%. U.S. imports from Mexico increased from $39.9 billion in 1993 to $294.2 billion in 2014, an increase of 637%.[36] In most sectors, NAFTA removed significant trade and investment barriers, ensured basic protections for NAFTA investors, and provided a mechanism for the settlement of disputes between investors and a NAFTA country. The agreement, however, included explicit country-specific exceptions and reservations. Under Chapter 6 of NAFTA, the Mexican government reserved to itself strategic activities, including investment and provisions in such activities, related to the exploration and exploitation of crude oil and natural gas.[37] Despite these exclusions from NAFTA, energy remains a central component of U.S.-Mexico trade, as discussed below.

Mexico Still a Top U.S. Oil Supplier

Mexico has been trading oil and natural gas with the United States since the turn of the last century. It is typically among the top three exporters of oil to the United States. Mexico's crude oil exports to the United States increased steadily in the 1980s and the 1990s, reaching a peak in 2004 and 1.6 million b/d. Since 2006, U.S. crude oil imports from Mexico have generally declined, reflecting Mexico's drop in production and rising internal demand. In 2014, the United States imported 842,000 b/d of crude oil from Mexico, behind Canada and Saudi Arabia.[38]

As shown in *Table 1*, Canada accounted for 34% of the dollar value of U.S. crude oil imports in 2014, followed by Saudi Arabia (18% of the total), and Mexico (11% of the total).

Table 1. U.S. Crude Petroleum Oil Imports in 2014
Total in Billions of U.S. Dollars

Country	Value	% of Total
Canada	83.2	34%
Saudi Arabia	44.9	18%
Mexico	27.7	11%
Venezuela	25.9	10%
Iraq	13.6	5%
Other Countries	51.8	21%
Total	247.0	—

Source: Compiled by CRS using USITC Interactive Tariff and Trade DataWeb: HTS number 2709 for crude petroleum oil.

The United States is the destination for approximately 71% of Mexico's oil exports, which arrive via tanker. Although Mexico has an extensive pipeline network that connects major production centers with domestic refineries and export terminals, it does not have any international oil pipeline connections. Exports leave the country via tanker from three Gulf Coast export terminals.[39] The majority of Mexico's crude oil exports are of the heavy Maya blend (approximately 82% of exports), while the lighter crude oil produced offshore is mostly retained for domestic consumption. Most of Mexico's crude oil exports will likely continue to be exported to the United States because of its close proximity and also because the U.S. Gulf Coast possesses the sophisticated refineries necessary to process the heavier Maya crude oil.

The leading U.S. import item from Mexico is crude petroleum oil.[40] The value of crude oil imports from Mexico in 2014 totaled $27.7 billion, nearly 30% higher than the value of automobile imports, the second leading import item. As shown in *Figure 6*, crude petroleum oil imports from Mexico have been considerably higher than other top imports from Mexico over the past 10 years. The drop in oil prices, however, caused the value of Mexican oil imports in 2014 to reach its lowest level since 2006, declining from a high of $39.6 billion in 2011 to $31.9 billion in 2013 and $27.7 billion in 2014.

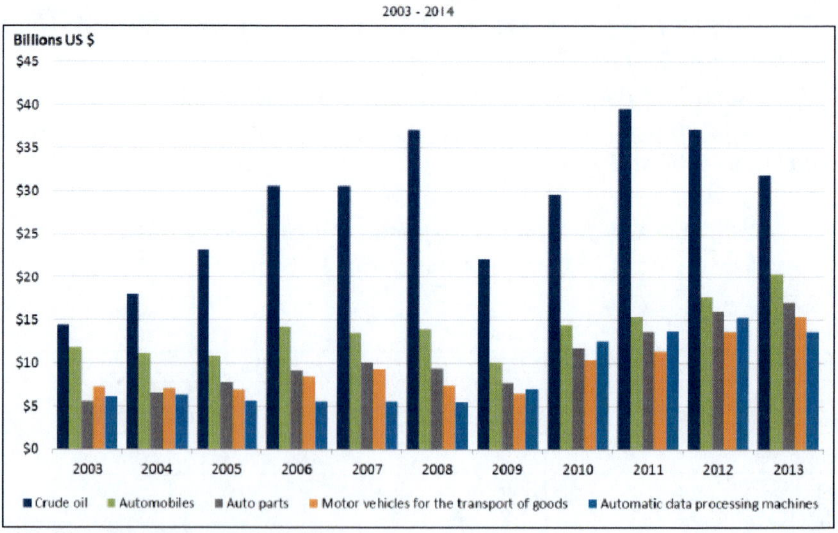

Source: Compiled by CRS using USITC Interactive Tariff and Trade DataWeb at http://dataweb.usitc.gov; HTS4-digit level.

Figure 6. Top 5 Imports from Mexico.

Although Mexico is one of the world's largest crude oil exporters, it is a net importer of refined petroleum products. In 2012, Mexico's imports of refined petroleum products from all countries totaled $29.6 billion. Mexico was the destination for 44% of U.S. exports of motor gasoline in 2013, although imports from the United States have declined since 2011.

Trade in Natural Gas

The United States has been Mexico's largest supplier of natural gas and Mexico continues to be a growing market for additional U.S. natural gas exports. As previously mentioned, Mexico's natural gas production has failed to keep pace with rising domestic demand making U.S. gas exports an important source of energy. The value of Mexico's natural gas imports increased from $995.7 million in 2007 to $2.5 billion in 2013.[41]

Mexico imported a total of 779 billion cubic feet (Bcf) of natural gas in 2012, out of which 620 Bcf came from the United States. The United States imports a very small amount of natural gas from Mexico. The surplus in natural gas trade with Mexico is expected to widen as recent supply and demand trends in both countries are expected to continue.[42] U.S.-Mexico trade in natural gas is done exclusively via pipeline.

Areas of U.S.-Mexico Energy Cooperation

In addition to the burgeoning energy trade between the United States and Mexico, energy cooperation has gradually risen to the top of the U.S.-Mexican political agenda as well. The United States and Mexico have been working on geothermal energy projects since the 1970s, but the possibility of expanding joint efforts to produce renewable energy sources, as well as conventional and unconventional hydrocarbons resources, has also entered the bilateral agenda.

Bilateral Framework on Clean Energy and Climate Change

On April 16, 2009, President Obama and then-Mexican President Calderón announced the Bilateral Framework on Clean Energy and Climate Change to jointly develop clean energy sources and encourage investment in climate-friendly technologies. Among others, its goals include enhancing renewable energy, combating climate change, and strengthening the reliability of cross-border electricity grids. Four bilateral meetings have thus far been held to advance the Framework. Since Mexico remains a top U.S. crude oil supplier and many of its untapped resources lie in deep waters in the Gulf of Mexico and in shale formations abutting the U.S. border, the countries want to ensure that Pemex (or other companies) develop those resources in an environmentally responsible way.

The U.S. and Mexican governments share a mutual interest in developing renewable energy sources, particularly those capable of serving rapidly growing population centers along the U.S.- Mexico border. As part of that effort, since 2011 the North American Development Bank has provided loans worth at least $677 million for projects related to wind and solar energy. The U.S. Agency for International Development (USAID) and Mexico have also expanded cooperation on environmental issues with the *Mexico Global Climate Change (GCC) Program*, a five-year, approximately $70 million program. The program seeks to help Mexico reduce emissions from deforestation, implement a low emissions development plan, and create a system for monitoring greenhouse gas emissions.

Although Mexico is trying to diversify its energy sources, it, like the United States, is likely to continue relying on oil and natural gas from traditional and unconventional (i.e., shale) sources. In the wake of the 2010 Deepwater Horizon spill in the Gulf of Mexico and amid concerns about the impact of hydraulic fracturing of shale oil in the United States, both governments have an interest in ensuring that hydrocarbons resources are developed in an environmentally responsible way. For example, as Pemex begins to partner with U.S. companies in the Gulf, it could benefit from participation in the Marine Well Containment Company that was created by U.S. companies to deal with spills. Should Pemex pursue contracts with U.S. companies who have years of expertise in hydraulic fracturing, they could complement the environmental and regulatory advice that Mexico is already receiving from the U.S. Departments of State and the Interior.

United States-Mexico Trans-Boundary Hydrocarbons Agreement

In 2012, the United States and Mexico signed an agreement known as the U.S.-Mexico Transboundary Hydrocarbons Agreement (the Agreement). The Agreement could mark the start of an energy partnership in an area of the Gulf of Mexico that the U.S. Department of the Interior estimates to contain as much as 172 million barrels of oil and 304 billion cubic feet of natural gas.

Although it concerns relatively little oil and natural gas, a main purpose of the partnership is to lift a moratorium on development in that region that had been in effect since 2000. In addition, the Agreement gives Pemex and U.S. companies options for jointly developing oil and gas reservoirs, referred to as "transboundary resources," that exist in areas straddling the marine border of both countries.

Prior to the expected expiration date for the moratorium, the Mexican and U.S. Congresses reviewed and accepted the Agreement.[43] Congress enacted legislation approving the Agreement on December 18, 2013, (H.J.Res. 59).[44] Now that the United States and Mexico have approved the Agreement, the moratorium is considered moot. The Department of the Interior's Bureau of Ocean Energy Management awarded the first leases subject to the Agreement in May 2014.[45]

Issues for Congress

Mexico's Economic Development

Mexico's long-term economic outlook depends largely on the energy sector. Economic growth in Mexico has been sluggish for several years, expanding just 2.4% in 2014 and 1.4% in 2013. Growth is forecast to drop slightly in 2015 to 2.1%. Since the Mexican government's reliance on oil income is very high, any decline in oil revenue has fiscal and economic implications. The decline in oil production combined with the drop in oil prices has resulted in lower export earnings, prompting foreign-exchange market volatility and causing the peso to depreciate.[46] Energy reform is the centerpiece of the President Peña Nieto administration's attempts to overhaul the economy, attract greater foreign investment, and generate more jobs. Numerous observers and government officials have laid out high expectations for the potential of Mexico's energy reforms to improve economic conditions. One study suggests that the reform "offers the prospect of enormous wealth generation over the next five years."[47]

According to some industry experts, the importance of energy reform in Mexico cannot be understated because of the additional opportunities for unconventional or shale oil and gas production.[48] While it is difficult to predict how increasing private participation in Mexico's oil and gas sectors may affect the country's economic development, skeptics see reason to doubt the government's positive predictions. Some argue that multinational companies and large Mexican conglomerates stand more to gain from the energy reform than the Mexican people.[49] Other critics question the government's claim that the reforms will create thousands of jobs, and, maintain that because Pemex is a bloated company with too many employees, it would likely shed workers as a result of the reform. Others are concerned that the oil revenue will be mishandled by corrupt Pemex or government officials rather than invested in strategic ways that will benefit the country as a whole.[50]

Impact on the U.S. Oil and Natural Gas Sectors and the Bilateral Energy Trade

The opening of Mexico's oil and natural gas sector to foreign investors poses significant changes in the U.S.-Mexico energy relationship that may have advantages and disadvantages for both sides. Reversing Mexico's production decline would add more oil to the global market and enhance U.S. energy security. Having a neighbor who is a growing oil producer to the south, as the United States has to the north with Canada, could provide a reliable supplier for the long term and it would also contribute to North American energy independence. Some predict that the region could produce surpluses of oil and gas in the coming decades, as well as expanded renewable energy: wind, solar, biofuels, and hydro-power.[51]

U.S. companies that are able to enter the Mexican upstream sector are likely to benefit from the opening of Mexican resources to foreign investment, depending upon the terms of the contracts that are offered. Significant opportunities for infrastructure development, oil services companies, and downstream industries are also likely to open up. This would be true for both the oil and natural gas sectors, but U.S. natural gas producers who export to Mexico might, over the long term, potentially lose some of their market.

U.S. Crude Oil Export Policy: The Case of Mexico[52]

U.S. crude oil exports generally are prohibited under U.S. law. The Energy Policy and Conservation Act of 1975 (P.L. 94-163, EPCA) directs the President to restrict the export of crude oil. There are certain cases where crude oil exports are permitted in statute: if it is shipped on the Trans-Alaska Pipeline, is of foreign origin, or is from the Strategic Petroleum Reserve if such export will directly result in import of refined products not otherwise available. EPCA and other statutes permit crude oil exports in circumstances where the President determines that such exports are in the national interest.

There have been some efforts to encourage the President to issue a national interest finding that would allow for unlimited crude oil exports from the United States to Mexico, as has been permitted for Canada since 1985. Under current laws and regulations, unlimited oil exports from the United States to Canada are allowed, although an application must be approved for each export transaction. As a result of this exemption, Canada is the primary destination for U.S. crude oil exports, with exports exceeding 570,000 bbl/d in May 2015.[53] The President could make a similar national interest determination for export to Mexico, and, arguably, such a determination would allow crude oil to be traded freely between the countries as economic conditions justified.

Congressional debate about potential changes to crude oil export restrictions is ongoing. Numerous economic studies from various government and industry organizations have been published. During the 114th Congress, there have been numerous hearings held by various House and Senate committees that have explored aspects of U.S. crude oil export policy. Furthermore, at least seven bills have been proposed during the 114th Congress, in addition to various Senate amendments that have been filed.[54] However, as of the date of this memo there has been neither committee nor floor votes on any of the proposed legislation or filed amendments.

Mexico-U.S. Oil Exchanges

Under the Export Administration Act of 1979,[55] the Bureau of Industry and Security (BIS) provides export licenses for crude oil exports under so-called short supply controls. Its licensing decisions are based on the statutory prohibitions and exemptions previously noted. However, it will review other applications to export crude oil on a case-by-case basis to determine whether they are in the national interest and consistent with the EPCA.

The BIS crude oil export regulations allow for exchanges with adjacent and non-adjacent foreign states as long as certain criteria are met. It is important to recognize that exchanges with adjacent and non-adjacent foreign states are distinct types of crude oil export transactions, and each has different criteria. Generally, crude oil export licenses for non-adjacent foreign state exchanges are much more difficult to obtain than those for adjacent foreign states. Adjacent foreign state exchanges could be approved based on "convenience and increased efficiency of transportation," terms that are not defined in the regulations. Crude oil exported as part of an adjacent foreign state transaction may not be reexported to another country.

In a move that could potentially enhance the efficiency of refinery utilization in both the United States and Mexico, the Mexican government proposed an exchange transaction whereby heavy, sour crude oil from Mexico would be exchanged for light, sweet crude oil from the United States. On January 8, 2015, the chief executive officer of Pemex announced that the company wants to increase Mexico's production of gasoline for domestic use by importing and refining light crude from the United States. [56] U.S. Commerce Secretary Penny Pritzker has confirmed that the United States is engaged in ongoing discussions with Mexican officials over whether to export light crude oil to Mexico.[57] Correspondence with the BIS at the Department of Commerce indicated that as of early July there are no new developments associated with the Mexico oil exchange application.[58] Timing for the license application decision is uncertain.

Pipeline Issues

Cross-border natural gas pipelines are authorized by the Federal Energy Regulatory Commission (FERC) in the United States. More than 20 natural gas pipelines linking the United States and Mexico are operating. Their approval by FERC has generally proceeded with little opposition, although concerns about environmental impacts from "induced" gas production in the United States may become a factor in the future. Between 2008 and 2013, U.S. pipelines to Mexico doubled their capacity. Current U.S. export capacity to Mexico is 4.7 billion cubic feet per day (bcf/d). By the end of 2015, that capacity should reach 7 bcf/d.[59] Private companies have, for the most part, not complained about the speed of the current pipeline approval process, although they have voiced concern about delays for specific projects.[60]

Unlike natural gas pipelines, cross-border oil pipelines require authorization from the U.S. Department of State in the form of a Presidential Permit. Although there are no crude oil pipelines between the United States and Mexico, there are two cross-border pipelines for refined petroleum products. The review and issuance of Presidential Permits by the State Department has become contentious in recent years, as illustrated by the protracted review of the proposed Keystone XL pipeline between the United States and Canada. A number of legislative proposals in the 113th Congress sought to modify the State Department's oil pipeline permitting authority to encourage pipeline development, but none have been enacted. The House approved a related bill on January 21, 2015 (H.R. 161).

Other Opportunities for Bilateral Energy Cooperation

As energy moves to the forefront of U.S.-Mexican relations, opportunities may exist for greater bilateral or trilateral (with Canada) energy cooperation. Those advocating a trilateral approach generally highlight the need for regulatory harmonization, regional planning, reduced investment and export restrictions, and strategic investments in new infrastructure.[61] Within the U.S.-Mexico context, analysts have urged the United States to offer more technical assistance to Mexico if it is requested.[62] Another area that could be expanded involves efforts to ensure that hydrocarbons resources are developed without unduly damaging the environment, possibly through collaboration between Mexican entities and U.S. federal or state regulatory entities. In terms of capacity-building, the University of Texas system has recently expanded educational exchanges and training opportunities for Mexicans working in the petroleum sector. Other U.S. universities could also follow suit, potentially with support from the Obama Administration's "100,000 Strong in the Americas" effort to boost educational exchanges.[63] Many others have also urged the United States and Mexico to work together to provide oil and natural gas resources to help reduce expensive energy costs in Central America as well.

TPP Negotiations

While the North American Free Trade Agreement (NAFTA) removed significant investment barriers and ensured basic protections for U.S., Canadian, and Mexican investors in other NAFTA countries, it did not open the Mexican energy sector to foreign investment.[64] The recent opening of Mexico's energy sector to foreign investors may have implications for the ongoing trade negotiations for a Trans-Pacific Partnership agreement (TPP).[65] The United States, Mexico and Canada are all participating in the talks and, while, a TPP is not likely to change NAFTA, an agreement may change the rules governing North American investment and trade. During an October 2014 visit to Washington, DC, Mexico's Economic Secretary Ildefonso Guajardo suggested that the TPP talks should reflect the opening of Mexico's energy sector to private investors and include updates in the investment provisions of NAFTA to deter trade and investment disputes among North American investors.[66]

OUTLOOK

Since taking office, President Peña Nieto has shepherded a number of significant constitutional reforms through the fractious Mexican Congress that had eluded the past two PAN Administrations. The most important of those reforms may be the energy reforms promulgated on December 20, 2013 and implemented by secondary laws signed on August 11, 2014 that allow for private participation in Mexico's oil and gas sector in ways not possible since the sector was nationalized in 1938. The recently-enacted energy reforms have the potential to boost energy production and improve economic competitiveness in Mexico, but implementing them in a meaningful way may prove difficult.

The next six months could prove critical for the success or failure of Mexico's energy reforms. Amidst strong opposition from the political left, the Mexican government will have to manage popular expectations about the benefits of the reforms, many of which may not be felt immediately. Pemex will need to restructure its workforce and investment priorities as it seeks to become a productive state enterprise that can compete with other companies. The National Hydrocarbons Commission (CNH) will need to complete the first round of public bidding in an efficient and transparent fashion and to offer contracts with terms that are attractive to international companies. At the same time, the executive will need to create strong regulators to oversee the hydrocarbons sector.

End Notes

[1] For background, see CRS Report R42917, Mexico: Background and U.S. Relations, by Clare Ribando Seelke, and CRS Report RL32934, U.S.-Mexico Economic Relations: Trends, Issues, and Implications, by M. Angeles Villarreal.

[2] The U.S.-Mexico High-Level Economic Dialogue (HLED) is a cabinet-led initiative launched in 2013 to advance economic cooperation. The White House, Office of the Press Secretary, "Joint Statement: United States-Mexico High Level Economic Dialogue," January 6, 2015. The North American energy ministers met for the first time on December 15, 2014. They

pledged to share data, develop best practices for unconventional hydrocarbons development, and modernize regional energy infrastructure.

[3] Lisa Viscidi and Paul Shortell, A Brighter Future for Mexico: The Promise and Challenge of Electricity Reform, Inter-American Dialogue, June 2014, http://www.thedialogue.org/uploads/IAD9603_MexicanEnergyFINAL.pdf.

[4] David H. Petraeus, Robert B. Zoellick, and Shannon K. O'Neil, North America: Time for a New Focus, Council on Foreign Relations, Independent Task Force Report No. 71, October 2014, available at http://www.cfr.org/northamerica/north-america/p33536. Hereinafter CFR, October 2014.

[5] Laurence Iliff, "Oil Majors Show Early Interest in Mexico Bidding Round," Wall Street Journal, January 20, 2015;

[6] U.S. Congress, House Committee on Foreign Affairs, Subcommittee on the Western Hemisphere, Pursuing North American Energy Independence: Mexico's Energy Reforms, 114th Cong., 1st sess., July 23, 2015.

[7] For an overview of Mexico's oil industry from 1901 through the 1970s, see George Grayson, The Politics of Mexican Oil (Pittsburgh, PA: University of Pittsburgh Press, 1980).

[8] Article 27 of Mexico's 1917 constitution gives the Mexican government exclusive legal authority to exploit, distribute, and process hydrocarbons in the country and states that the government may not, per the regulatory law, grant private concessions for their exploitation. Article 28 establishes petroleum and other hydrocarbons as strategic sectors over which the government (public sector) exerts total control.

[9] Adam Thompson, "Rusty Wheels of Pemex Require Much Oiling," Financial Times, April 3, 2013.

[10] Reforms passed in 2006 allowed Pemex to fund a portion of its capital investments with earnings rather than through the issuance of new debt. Ognen Stojanovski, "Handcuffed: an Assessment of Pemex's Performance and Strategy," in Oil and Governance: State-owned Enterprises and the World Energy Supply, ed. David G. Victor, David R. Hults, and Mark C. Thurber (Cambridge University Press, Cambridge: UK, 2012).

[11] Thompson, op. cit.

[12] Patrick Corcoran, "Oil Theft Is Big Business for Mexican Gangs," InSight Organized Crime in the Americas, March 20, 2012.

[13] For information on the reforms in English, see http://reformas.gob.mx/en/reforms.

[14] Under this scenario, private companies receive a percentage of the remaining revenue earned once exploration and production costs have been recovered, but would not own a share of production.

[15] Upstream refers to the exploration, development, and production phases of oil and natural gas production. Midstream refers to the transportation of the resource, and downstream refers to the refining and marketing of the resource. Natural gas does not need to be refined, so downstream in that sector relates to the marketing of natural gas.

[16] President Calderón originally proposed measures that would have allowed Pemex to enter into joint ventures with foreign companies in exploration and production, and permitted private companies to build and operate refineries, pipelines, and storage facilities in Mexico. Calderón's proposal prompted strong resistance from the PRD and was significantly watered down by the PRI in the Mexican Congress. Nevertheless, the final legislation brought private sector experts into Pemex's management structure, created an independent National Hydrocarbons Commission (CNH) to advise the company, and added greater flexibility to its procurement and investment processes. Most significantly, the 2008 reforms permit Pemex to create incentive-based service contracts with private companies.

[17] David L. Goldwyn, Mexico Rising: Comprehensive Energy Reform at Last? Atlantic Council, December 2013, http://www.atlanticcouncil.org/images/publications/Mexico_Rising.pdf.

[18] The reforms were approved by the Mexican Senate on December 11, 2013 (95-28) and the Chamber of Deputies on December 12, 2013 (354-134).

[19] This section draws from David L. Goldwyn et al., Mexico's Energy Reform: Ready to Launch, Atlantic Council, August 2014, available at http://www.atlanticcouncil.org/ publications/ reports/mexico-s-energy-reform-ready-to-launch.

[20] Atlantic Council Fact Sheet, available at http://www.atlanticcouncil.org/images/files/ Energy_Reform_Fact_Sheet.pdf.

[21] Ibid. Pedro Valenzuela and Duncan Wood, Mexico's Energy Reform: Entering the Final Phase - The Expert Take, Woodrow Wilson Center's Mexico Institute, August 13, 2014.

[22] Eric Martin and Brendan Case, "Mexico Cuts Spending Saying Oil Price May Stay Low for Years," Bloomberg, January 30, 2015.

[23] Robert Perkins, "Pemex Sees No Price Constraints to Reforms," Platts Oilgram News, October 31, 2014.

[24] "First Phase of Mexico's 'Round 1' Oil Concessions Disappoints," Latin News Daily, July 16, 2015.

[25] BP Statistical Review of World Energy, June 2014, http://www.bp.com/en/global/corporate/ about-bp/energyeconomics/statistical-review-of-world-energy.html.

[26] Tight oil refers to oil that is trapped in impermeable formations, such as shale or sedimentary rocks, and requires artificial fractures to allow the hydrocarbons to flow.

[27] U.S. Energy Information Administration (EIA), Technically Recoverable Shale Oil and Shale Gas Resources: An Assessment of 137 Shale Formations in 41 Countries Outside the United States, Washington, DC, June 10, 2013, p. 10, http://www.eia. gov/analysis/ studies/worlds halegas/. Hereinafter: EIA, June 2013.

[28] This paragraph draws from: EIA, Country Analysis: Mexico, April 24, 2014, available at http://www.eia.gov/ countries/cab.cfm?fips=MX.

[29] EIA, "Energy Reform Could Increase Mexico's Long-Term Oil Production by 75%," press release, August 25, 2014.

[30] Electronic correspondence with U.S. Federal Energy Regulatory Commission (FERC), October 3, 2014.

[31] EIA, June 2013.

[32] Kathryn Haahr, Addressing the Concerns of the Oil Industry: Security Challenges in Northeastern Mexico and Government Responses, Wilson Center: Mexico Institute, January 2015, available at http://www.wilsoncenter.org/ publication/addressing-the-concerns-the-oil-industry-security-challenges-northeastern-mexico-and.

[33] EIA, April 2014.

[34] Justin Jacobs, "Refining Woes in Latin America," Petroleum Economist, October 2013.

[35] "Mexico's Pemex to Incur $7.7 Billion Refining Loss in 2013," Fox News Latino, October 23, 2013.

[36] Based on data from the U.S. International Trade Commission (USITC) Interactive Tariff and Trade DataWeb using Harmonized Tariff Schedule (HTS) at the 4-digit level. For more information on NAFTA and its effects on trade, see CRS Report R42965, The North American Free Trade Agreement (NAFTA), by M. Angeles Villarreal and Ian F. Fergusson.

[37] Chapter 6 of NAFTA applies to energy and basic petrochemicals. It reserves most activities in Mexico's energy sector to the Mexican state.

[38] EIA, Mexico Country Report, April 24, 2014.

[39] EIA, April 24, 2014.

[40] USITC Trade DataWeb, using HTS code 2709 for crude petroleum oil.

[41] Secretaría de Energía de México, http://www.sener.gob.mx.

[42] EIA, April 2014.

[43] The Mexican Senate reviewed and accepted the agreement in April 2012. On June 27, 2013, the U.S. House of Representatives passed H.R. 1613, the Outer Continental Shelf Transboundary Hydrocarbon Agreements Authorization Act (H.Rept. 113-101). House activity featured signs of a persistent policy divide between "pro-drilling" arguments to accelerate energy production and "anti-drilling" arguments to maintain the moratorium in order to provide time for fiscal, safety, and environmental issues to be addressed. On October 12, 2013, the Senate passed S. 812, to allow the Secretary of the Interior to implement the Agreement.

[44] Title III of the Bipartisan Budget Act of 2013 (H.J.Res. 59) gives congressional approval of the Agreement and requires implementation planning by the Secretary of the Interior. For details about implementation plans to date, see http://www.bsee.gov/uploadedFiles/BSEE/Newsroom/Publications_Library/BOEM-BSEE%20MOU%20and%20Attachment%20A%20re%20Impementation%20of%20US-Mexico%20TBA%20(FINAL%20signed%207-17-2014).pdf.

[45] "BOEM awards leases in US-Mexico transboundary area to ExxonMobil," Oil & Gas Journal, May 30, 2014.

[46] IHS Connect, Country Outlook: Economic - Mexico, Country Reports, June 24, 2015.

[47] Thomas Tunstall, Ph.D., Javier Oyakawa, M.S., M.Sc., and Alejandra Bueno, et al., Economic Impact and Legal Analysis of the Shale Oil and Gas Activities in Mexico, UTSA, UANL, AEM, Wilson Center, Preliminary Report, May 2015.

[48] Ibid.

[49] "Richard Fausset, "Tons of Thousands Protest Mexican Oil Reforms," Los Angeles Times, September 8, 2013.

[50] Enrique Krauze, "Mexico's Theology of Oil," New York Times, November 1, 2013.

[51] CFR, 2014.

[52] This draws from CRS Report R43442, U.S. Crude Oil Export Policy: Background and Considerations, by Phillip Brown et al.

[53] Oilgram Price Report, "Canadian Imports of U.S. Crude Hit Fresh Highs in May," July 9, 2015.

[54] Proposed legislation in the 114th includes H.R. 156, H.R. 702, H.R. 1487, H.R. 2369, S. 791, S. 1312, and S. 1372.

[55] The EAA is currently expired, however, its provisions and those of the regulations implementing it, the Export Administration Regulations (EAR), are maintained through a declaration of national emergency and invocation of the International Economic Emergency Powers Act (IEEPA, P.L. 95-223).

[56] Voice of America, Mexico Proposes Historic Crude Oil Swap with US, January 8, 2015.

[57] David Alire Garcia and Christine Murray, "U.S. Discussing Possible Light Crude Exports to Mexico: Pritzker," Reuters, January 6, 2015.

[58] Email correspondence with the Bureau of Industry and Security, July 8, 2015.

[59] Electronic correspondence with U.S. Federal Energy Regulatory Commission (FERC), October 3, 2014.

[60] On June 24, 2014, the House approved legislation (H.R. 3301) that seeks to ensure the continued development of natural gas infrastructure and exports to Mexico. Those provisions were integrated into H.R. 2, which the House passed on September 18, 2014.

[61] CFR, 2014.

[62] Goldwyn, August 2014.

[63] For information, see http://www.100kstrongamericas.org/about.

[64] For more information on NAFTA, see CRS Report R42965, The North American Free Trade Agreement (NAFTA), by M. Angeles Villarreal and Ian F. Fergusson.

[65] For more information on the TPP, see CRS Report R42694, The Trans-Pacific Partnership (TPP) Negotiations and Issues for Congress, coordinated by Ian F. Fergusson. The 12 countries involved in the Trans-Pacific Partnership (TPP) negotiations are the United States, Australia, Brunei, Canada, Chile, Japan, Malaysia, Mexico, New Zealand, Peru, Singapore, and Vietnam.

[66] "Trade Talks Must Reflect Mexico's Energy Reforms, Guajardo Says," Bloomberg Bureau of National Affairs, October 23, 2014.

In: Mexico's Oil and Gas Industry
Editor: Neil Becker

ISBN: 978-1-63483-846-7

Chapter 2

TESTIMONY OF CARLOS PASCUAL, SENIOR VICE PRESIDENT, IHS INC. HEARING ON "PURSUING NORTH AMERICAN ENERGY INDEPENDENCE: MEXICO'S ENERGY REFORMS"*

Chairman Duncan, Ranking Member Sires, and members of the Committee, I appreciate the opportunity to testify before you on Mexico's Energy Reforms and their significance for North America.

Since 2012, Mexico has embarked on an historic opening of its energy sector to allow private investment and competition in the production, transit and sale of oil, gas and electricity, and in the coming years in retail markets as well. On July 15, Mexico completed the first tender since 1938 for the sale of hydrocarbon assets. Even though the results did not meet expectations, it formalized a process of opening the energy sector to private investment – and with that, the benefits that will eventually ensue from infusions of capital and technology. Mexico will benefit from these reforms, but so will American businesses and workers. North America can improve its energy security and play a more profound role in stabilizing energy markets regionally and globally. I will address these issues in my testimony.

I appear before you in my capacity as Senior Vice President for IHS. IHS is a global consultancy that specializes in energy, capital-intensive industries,

---

* This is an edited, reformatted and augmented version of testimony presented July 23, 2015 before the House Foreign Affairs Committee, Subcommittee on Western Hemisphere.

data and analysis with a worldwide presence. While my responsibilities are global, I am based in Mexico, and have had the opportunity to engage many of the key players in Mexico's energy reforms. I previously served as the Coordinator for International Energy Affairs and Special Envoy on Energy at the State Department. I am associated with Columbia University as a Fellow at the Center on Global Energy Policy. I served as the U.S. Ambassador to Mexico from 2009 to 2011.

MEXICO'S ENERGY REFORMS

Although Mexico nationalized its hydrocarbons sector in 1938, energy cooperation between the United States and Mexico has a long history. U.S. oil investor Henry Clay Pierce helped finance the start of Mexican oil production in 1901. American oil entrepreneur Edward Dohenny, one of the pioneers of the oil industry in California, founded the Pan American Petroleum and Transit Company in Mexico and drilled the Cerro Azul No. 4 well in Mexico, which became the world's largest producing oil well at the time. By the early 1920s, Mexico became the world's largest exporter of oil and the second largest producer after the United States.

The House of Representatives played a key role in resuscitating that tradition of cooperation by championing the passage of the U.S.-Mexico Transboundary Hydrocarbons Agreement in 2013, which has made possible the development of transboundary reserves that require the cooperation of both the United States and Mexico. The Agreement creates a framework for U.S. and Mexican companies to work together on the joint exploration and production of reservoirs that cross our maritime boundary anywhere in the Gulf of Mexico. As a result, it provides a foundation for increasing energy supplies that will benefit both economies and it promotes cooperation between our two countries on safety and environmental regulation, including providing for joint inspection and enforcement of activity under the Agreement.

Although Mexican oil production grew steadily in the 1900s, as of 1938 the Mexican constitution prohibited private investment, and as a result deprived Mexico of revolutionary progress in energy technologies. Production peaked in 2004 at about 3.6 million barrels per day (bpd), and since then has declined to current levels of about 2.35 million bpd – a 35 percent decline. This decline, contrasted with the general international practice of allowing international companies to participate in oil and gas production around the world, influenced the Mexican government under President Enrique Peña

Nieto to make energy reform a core pillar of its policy agenda when it took office in December 2012. Equally important was the recognition that a healthy energy position was necessary to support economic growth and create opportunities for Mexican workers, and that such a position required opening up the industry. The pace, breadth and depth of implementing these energy reforms has been unprecedented.

In December 2013, within a year of taking office, Mexico passed a constitutional reform to allow private investment and competition in every aspect of its energy sector. This included the support of two thirds of the Mexican House of Representatives and Senate, and a majority of the 32 state legislatures. On August 12, 2014, Mexico passed the secondary legislation needed to implement reforms related to oil and gas production, pipelines, private competition in the power sector, and competitive retail sales. The very next day, Mexico announced the results of the so-called Round 0, allocating 83% of the proven and probable oil reserves to the state oil company PEMEX, but in so doing making clear that this was the basis to end PEMEX's role as a government agency and transform it into a productive enterprise that must survive on its own resources. In December 2014, just two years after it took office, the Mexican government announced its first round of international tenders – what they call Ronda 1 – with five phases that will consider the sale of about 160 fields over the course of a year. I will come back to the implementation of Round 1 and future prospects later in the testimony.

In the power sector, by end of August 2014 Mexico created an independent systems operator for a new electricity market called the National Center for the Control of Energy (CENACE). In doing so, it created the foundation for competition with the monopoly held by the Federal Electricity Commission (CFE). CFE could contract with independent power producers (IPPs) for power supply, but under the new market, these IPPs will be also able to bid against CFE to supply electricity to the power market, breaking the government's monopoly control over electricity supply. By February 2015, the Mexican Government published the draft new market rules for the power sector. In September it will conduct a live test of the power market, and the new power market and its associated elements for a capacity market, clean energy certificates, and transmission and distribution will come into force in January 2016.

Private investment is also advancing in pipelines and refineries, and will be open for retail sales in 2016. Opportunities will expand further when subsidies are phased out on liquefied petroleum gas (LPG) in 2016 and for gasoline in 2018. Both CFE and PEMEX have completed tenders to expand the national gas pipeline system and to complete new interconnections with the United States. This expansion of the pipeline system will bring U.S. natural gas to Mexico, help reduce household and industrial costs, and increase North American competitiveness. By the end of 2015, the Mexican Energy Regulatory Commission will complete new regulations that will allow investments in existing and new refineries, either as a partner or in competition with PEMEX.

NORTH AMERICA'S INTERESTS

The strategic and commercial significance of these reforms for the United States and North America cuts across the energy sector and beyond.

In hydrocarbons production, the completion of these reforms gives the United States, Canada and Mexico an opportunity to make North America a new foundation for global energy security. On November 27, 2014, OPEC demonstrated that, today, it would not act as a bloc to adjust production and affect energy prices. Instead, Saudi Arabia, Kuwait and UAE decided to protect their global market share in oil, and to rely on market forces to rebalance supply and demand to the disadvantage of high cost producers.[1] North America can contribute to filling the space left by OPEC to help shape global oil market conditions that drive stable and sustained economic growth. The U.S. Energy Information Administration has projected that the United States, in a high case scenario, could increase crude production from current levels of 9.3 million bpd to 13.3 million bpd in 2020. The National Energy Board of Canada's high forecast estimates that Canada could increase its production from 3.6 million bpd in 2015 to 4.3 million bpd in 2020. Mexico now has the chance to enter this club of rising production as its energy reforms attract investment. North America will not act like an oil cartel. Rather, it represents three democratic and market-oriented states establishing a reliable base of production that will set standards in international conduct and transparency in energy development and trade that can influence the global industry. But to achieve this new foundation, the United States must be an exporter, a key issue addressed below in the policy recommendations.

Mexico's new focus on extending gas pipelines has created business opportunities for U.S. companies and investors. In one pipeline extending from Arizona to Mazatlan on the Mexican Pacific coast, El Paso Natural Gas (a Kinder Morgan subsidiary) won the contract to build the U.S. portion. For the Los Ramones II pipeline, which completes a route from Agua Dulce,

Texas to Aguascalientes and Queretaro in Mexico, the American investment firms Blackrock and First Reserve have invested $900 million to acquire 45% ownership. On June 22, 2015, CFE announced tenders for 24 energy infrastructure projects estimated at $10 billion in investment, including a $3.1 billion subsea pipeline from Brownsville to the Mexican port of Tuxpan and seven additional pipeline projects. U.S. companies have comparative advantages to participate in construction and finance. The extension of natural gas pipelines and lower electricity prices directly benefit companies like GM, Ford, Chrysler, Boeing, GE, Intel and other major American companies invested in Mexico, operating integrated production lines and global export platforms that have made their American products more competitive in the United States and globally.

Similar investment patterns in the power sector and refineries are possible as final regulations and market rules come into effect. CFE has launched a major program to invest in natural gas power, creating opportunities for suppliers of generation equipment, but also leaving much of the renewable energy sector open to private investors. Mexico imports 40% of its gasoline and has huge and urgent demand to upgrade existing refineries to meet environmental standards and expand supply. All of these openings in the energy sector will certainly drive Mexico's economic growth and competitiveness, but they represent a huge strategic and business opportunity for the United States. To realize that potential, Mexico and its investment partners will have to manage difficult market challenges.

Ronda 1 and Global Oil Markets

Since Mexico amended its constitution to allow private investment in energy, the price of the Brent benchmark price for crude oil has dropped from a range of $105- 110 per barrel to about $50-60 per barrel over recent months. This hearing is not the forum to address the reasons for this price collapse, but there is no doubt that it affects the implementation of Mexico's reforms. Lower prices are forcing oil companies to reduce capital expenditures,[2] financial institutions to cut investments in independent producers, and oil producing countries to demonstrate that their fiscal terms compete with the best international alternatives. Mexico has a disadvantage: international players do not have first-hand knowledge of Mexican assets and how to operate in Mexico. Simply put, to compete effectively, Mexico has to make completely clear that its assets, contracts, fiscal terms, and local business and security environments offer investment returns that attract production and capital to Mexico at a time when the industry is cutting expenditures and costs.

Mexico has announced plans for five phases of bids under Round 1 of its hydrocarbons tender process: (1) shallow water exploration (14 blocks), (2) shallow water production (5 blocks), (3) onshore (26 blocks), (4) deep water (13 blocks) and extra heavy oil (11 blocks), and (5) unconventional fields (61 blocks) and Chicontepec (a massive, complex onshore field,12 blocks). Bids for phase 1 were opened on July 15. Bids will be opened for phases 2 and 3 on, respectively, September 30 and December 15. The timelines for the other phases have not yet been announced.

On July 15, Mexico saw the impact of the difficult business climate in global oil production with the disappointing interest in Phase 1. 39 companies paid for data to investigate the blocks offered in Phase 1, 34 companies prequalified to compete, but on July 15 there were only 9 bidders, with 4 of the bids coming from consortia. Only two out of fourteen blocks were awarded. Six blocks received offers, but five were under the minimum bids set by the government and were not awarded, and one lost to another bidder. The consortium formed by Sierra Oil & Gas (a newly formed Mexican company), Talos Energy, and Premier Oil won both blocks. Talos is a Houston-based company.

Mexican officials will spend time investigating and applying lessons from Phase 1, either to contract terms for phases 2 and 3, but most likely for a deep-water tender that has yet to be announced. Some key issues might include:

- Field offerings were small, and perhaps not of high interest to the larger international companies.
- Contracts were offered for four years with a two-year extension, but some companies may have wanted longer contract terms to perform more extensive exploration, such as whether there might be complex pre-salt formations that could be exploited at deeper levels.
- Government minimum bids may have been influenced by historic PEMEX production costs, which may be lower than the costs estimated by potential investors.
- Fiscal terms may not have met investor requirements to mobilize capital given increased pressure from low international oil prices to cut costs and capital expenditures.

The important point is that all of these issues can be addressed. Mexico took an historic first step, demonstrating the potential for an important combination of international oil partners, American capital, and a start-up Mexican private industry. There will be more issues to address, such as the capacity of the Mexican service sector, whether other Mexican private companies can secure international partners, the role PEMEX will play in future bid rounds, and pipeline and storage capacity. However, with the sector now open for investment, these potential issues can be turned into business opportunities. If the Mexican government and outside investors take that perspective, these difficult first moments could help consolidate the profound need for sensitivity to international markets and competitiveness that Mexico needs to demonstrate to fulfill its aspirations.

TRANSFORMING THE POWER SECTOR

Less attention has been focused on electricity than on oil production, but the transformation of the power sector could be the lynchpin that sustains political support for energy reform. By the end of the President Peña Nieto's Administration, successful implementation of its oil tenders should allow Mexico to demonstrate contracts for tens of billions of dollars in new investment, but those investments will take time to halt and reverse current production declines. In the power sector, the fast expansion of U.S. natural gas imports has already allowed CFE to lower Mexican electricity prices by 27% for industry, and between 2-11% for households. Investments in natural gas pipelines from the United States will help Mexico sustain these major and tangible gains that benefit consumers and industry and help create jobs. President Peña Nieto retains a majority in parliament after mid-term elections. Polls suggest that the next government and parliament after 2018 will still support energy reform. But the best way to secure that support is if the Mexican people benefit tangibly from the results – and here the power sector is key.

As in oil and gas, there will be implementation challenges as Mexico overhauls every aspect of its power sector. It will face a conundrum many states have seen in the United States: as competition drives down prices, how do you create an incentive to invest in capacity? As competition increases with legacy utilities and takes away parts of their base market, how do you manage stranded assets that are not competitive, must be closed, and workers laid off? As clean energy targets are established, what are the cost implications and how are intermittent renewables integrated with base-load generation requirements? It is inevitable that Mexico will face these questions and need industry partnerships and technical support to avoid mistakes where it can, and adjust quickly when needed.

Both CFE and the Mexican government's National Program for the Electricity Sector make clear that natural gas will underpin the Mexican power sector and CFE's investments through 2020. Yet Mexico has also made very aggressive commitments on clean (zero greenhouse gas emission) energy: 35% of all generation by 2024, with emissions peaking in 2026. It will reduce GHG emissions between 25-40% by 2030 relative to business as usual, depending on market conditions. To meet these targets, Mexico will conduct long-term auctions for new capacity. With CFE focused on the gas sector, the renewables sector will offer a huge target for private investment, where the United States can be a major supplier and investor.

Policy Implications

From this discussion, let me highlight five areas for policy actions that might be considered by this Committee that would enhance the future energy position of North America.

Oil Exports. I indicated earlier the strategic benefits of cooperation among Mexico, Canada, and the United States to make North America a foundation for energy security – for our own region, and with a stabilizing impact globally. North American self-reliance on energy would be a critical point of flexibility and leverage in policy discussions with countries in Asia and the Middle East. North America will still depend on global oil prices – we cannot disconnect from global markets – but we are less vulnerable to interruptions of energy. To fully realize the benefits of our energy abundance, we should also have the opportunity to export oil. Other Congressional committees have addressed this issue. IHS has completed two exhaustive studies3 that demonstrate that lifting the ban on oil exports will bring major benefits across the United States between 2016-2030:

- $86 billion in additional GDP,
- about 400,000 new jobs annually,
- 25% higher pay for workers in the energy industry supply chain – an additional $158 per household, and
- $1.3 trillion in federal, state and municipal revenue from corporate and personal taxes.

The strategic benefits for North America only accentuate the benefits we would see from lifting the export ban on the U.S. economy.

2. North American Energy Forum. Before the passage of its energy reforms, Mexico could not address energy seriously in its annual North American Leader Summits and its parliamentary exchanges with the United States. That can now change and be underpinned by a North American Energy Forum that would include the energy ministers of all three countries, and also establish a forum for private company engagement. To be effective, it would create committees or working groups that facilitate public-private engagement. In oil and gas, companies and governments could cooperate on sharing best practices on issues from capturing methane emissions, to hydraulic fracturing, to transparency in contracting and payments. Such a forum would generate inputs for policy measures, and perhaps financing opportunities. It would make cooperation among energy leaders in government, legislatures, and the private sector more concrete and presumably more effective.

3. The Caribbean. Since 2005, Venezuela has subsidized oil for the Caribbean states through an initiative called Petrocaribe as a means to increase its relevance and influence and bolster its opposition and hostility to the United States. The global collapse in oil prices has deepened what was already a severe financial crisis in Venezuela, compromising its ability to sustain these subsidies. The loss of these subsidies will create economic hardship for many of the Caribbean nations. But with lower oil prices, the Caribbean states have a chance to find competitive sources of supply. There is currently an active dialogue among the United States, Canada and Mexico through the North American Leaders Forum to open new sources of supply or financing for alternative forms of power that reduce the Caribbean's dependence on high-priced diesel. The openings in Mexico for private investment and international engagement will make it easier to make Mexico a partner in such endeavors. To have an impact, the United States will need to invest in technical assistance and financing, potentially through the Overseas Private Investment Corporation, to bring alternatives to Petrocaribe within commercial reach.

4. Central America. A power line now exists from Mexico to Panama that has begun to facilitate electricity trade across Central America. The Inter-American Development Bank has helped finance infrastructure and technical assistance costs. The United States has provided technical assistance to help develop the market rules and conditions for private investment. The question of power access and costs is a critical underlying factor in the Central American economies, which may pay 20-50% more for power than we would pay in the United States. But just as important, the lack of reliability affects

business interests, job creation, and educational opportunities – all factors central to combating organized crime and illegal migration from Central America to Mexico and the United States. Mexico is already an electrical power source for Central America, and it could provide additional supplies at competitive prices that would reduce costs and increase reliability in the region. There is also a joint effort among Mexico, Guatemala, El Salvador and Honduras to build a natural gas pipeline from Mexico at an estimated cost of $1.2 billion. The pipeline would introduce another lower cost energy and fuel source. This Committee is well placed to encourage the Administration to engage aggressively with Mexico on energy supply to Central America, and to help secure the resources for technical assistance through the State Department and USAID, and for investment through power market in Central America contribute to regional stability.

5. Mexican Power. We have already discussed the complex challenges that face the Mexican power sector. The United States has skills and experience to contribute, perhaps most usefully through energy partnerships between utilities, and through targeted technical assistance. Such programs exist in USAID, the State Department and the Department of Energy. There is now a political opening that did not exist before Mexico's energy reforms to target and expand these in ways that will support private investment.

Mr. Chairman, I congratulate the Committee for its leadership in addressing the potential that Mexico's energy reforms offer strategically and commercially – not just to Mexico, but to the United States and North America. In the energy field, no other region has today what North America can offer: energy abundance and technology leadership across three democratic states and market economies, with huge consumer markets, financing potential, and a global reach and influence. The successful implementation of Mexico's reforms is critical to make the most of these opportunities. The global oil price collapse has made the reform challenge more complex for Mexico. The correct response is to learn from these market conditions, adapt, and use these lessons to capture the potential for investment, production and trade. Thank you for the opportunity to address the committee. I look forward to your questions.

End Notes

[1] IHS, Oil Change: A World Without OPEC as We Knew It (February 2015)

[2] IHS Upstream Spending Report – 1Q 2015. Global exploration and production (E&P) spending is expected to drop by 16.6% year on year in 2015, followed by a further 12.6% decline in 2016.

[3] See IHS, US Crude Oil Export Decision: Assessing the impact of the export ban and free trade on the US economy (2014), and IHS, Unleashing the Supply Chain: Assessing the economic impact of a US crude oil free trade policy (2015).

In: Mexico's Oil and Gas Industry
Editor: Neil Becker
ISBN: 978-1-63483-846-7

Chapter 3

TESTIMONY OF DR. THOMAS TUNSTALL, RESEARCH DIRECTOR, INSTITUTE FOR ECONOMIC DEVELOPMENT, THE UNIVERSITY OF TEXAS AT SAN ANTONIO. HEARING ON "PURSUING NORTH AMERICAN ENERGY INDEPENDENCE: MEXICO'S ENERGY REFORMS"*

Good afternoon. I would like to thank Chairman Duncan and the members of the subcommittee for extending me the privilege to testify here today. My name is Thomas Tunstall. I am the research director for the University of Texas at San Antonio Institute for Economic Development. Our institute has undertaken extensive research to-date on the Eagle Ford Shale in South Texas, specifically addressing issues dealing with the economic impact and sustainable community development. In September of last year, we completed our fourth economic impact report on the Eagle Ford in Texas. In that report, for the calendar year 2013, we estimated that the Eagle Ford Shale generated $87 billion in economic impact, supporting over 150,000 full-time jobs. While the recent decrease in oil prices has resulted in some job loss, the activity in the Eagle Ford in South Texas continues to have a significant impact on the region – one of the traditionally poorest areas of the state if not the entire

* This is an edited, reformatted and augmented version of testimony presented July 23, 2015 before the House Foreign Affairs Committee, Subcommittee on Western Hemisphere.

country. Prior to the ramp-up of activity in the Eagle Ford, many of the counties in the region were losing population. This unexpected windfall has given communities in the area an opportunity to build a foundation for long-term sustainable economic and community development.

The Eagle Ford is one of the more interesting shale formations that have been extensively explored to-date. The shale is noteworthy because it contains significant quantities of oil and gas, as well as condensate, also known as wet gas or ultralight crude oil. The Eagle Ford is also arguably the most productive shale formation for oil production in the United States or anywhere else in the world so far. And the while the same shale field is called the Burgos Basin across the Rio Grande border in Mexico, the Eagle Ford and other formations continue well into Mexico, over to Monterrey, and then along the Gulf Coast as far south as Vera Cruz.

It is worthwhile to note that Mexico's oil production peaked in 2004, and has been declining steadily since then. In fact, were it not for energy reform in Mexico, the country would likely have been facing the prospect of becoming an oil importer in the next few years. Mexico already imports substantial quantities of natural gas from the U.S. – over 650 billion cubic feet annually in 2013 and 2014, with even greater quantities expected in the next few years by way of additional pipeline capacity that has recently been developed. At the same time, shale gas reserves alone in Mexico are estimated to be 545 *trillion* cubic feet.

Mexico's energy reform consists of several types of fields and blocks that will be auctioned off in the coming months. These blocks include deep-water fields, shallow water fields, onshore conventional fields, and shale fields. Our research at the University of Texas at San Antonio has focused primarily on the shale field opportunities in the Mexican states of Coahuila, Nuevo Leon, Tamaulipas and Vera Cruz.

While the experience in the Eagle Ford in Texas can be a useful point of reference for the opportunities for Mexico's shale, there are some important differences to note. Although there have been infrastructure challenges in South Texas, most were overcome relatively quickly. In Mexico, the infrastructure challenges will be much more significant. In order to address some of the key issues at hand, I would like to highlight four policy prescriptions that I believe would encourage the development of the Mexican energy industry, increase trade activity between the U.S. and Mexico, and bolster North American energy security.

POLICY PRESCRIPTIONS

1. Expedite Oil Swap Approval

Most of the refineries along the Gulf Coast in the U.S. were optimized to process heavier crude oil that the U.S. expected to be importing from OPEC countries or Canada. However, because of the unexpected increase in light crude oil production from shale, the refineries are processing a type of oil that they did not expect to receive. Much of Mexico's production tends to be heaver crudes. And in addition, much of Mexico's refining capacity is better equipped to process lighter crudes of the type coming from U.S. shale fields. Both from an optimization as well as energy security standpoint, it makes sense to encourage the Department of Commerce to approve the oil swap arrangements with Mexico as soon as possible.

2. Streamline Border-Crossing Process for Key Industry Employees in Order to Facilitate Workforce Development

The bulk of worldwide expertise dealing with unconventional extraction techniques resides in the United States, especially in Texas. While the ultimate success of energy reform still depends on the implementation, Mexico's proximity to Texas puts the country in a good position to benefit from the technologies and techniques developed in the United States. However, in order for a suitable workforce to be developed in Mexico, it is often necessary for employees to be trained by experienced staff on working rigs located in the United States. Typically, Mexican nationals sponsored by exploration and production companies will make several training visits to the U.S., sometimes daily. Experienced U.S. workers will also increasingly be making trips to Mexico to supervise operations. While such border crossings would be expected to be a routine procedure, experience to-date suggests that significant delays of hours at a time are not uncommon, which burden operators with unnecessary costs and delays. Developing a streamlined process for worker knowledge-transfer from the U.S. to Mexico will be an important step to ensure the ultimate success of Mexican energy reform implementation.

3. Supply Chain Logistics between the U.S. and Mexico

Unconventional techniques require a wide variety of personnel and equipment to complete a shale well. The logistical infrastructure in northern Mexico in terms of roads, housing, workforce, medical facilities, rail, telecommunications and pipelines is significantly undeveloped. In order to successfully develop the shale opportunities in Mexico, the federal and state governments there (perhaps in coordination with the U.S.) must make a commitment to invest in the infrastructure that will be necessary to support oil and gas development, as well as provide a foundation for other types of industry in the future. Developing the supply chain network that will be required represents a significant opportunity to increase the involvement in Mexico for small-medium enterprises in Texas and other oil producing states.

4. Crude Oil Export

U.S. hydrocarbon export policy is inconsistent. Currently, U.S. law allows for unlimited, unrestricted export of refined products such as jet fuel, gasoline and diesel fuel. Further, natural gas can be readily exported to any country with which the U.S. maintains a free trade agreement, such as Mexico and Canada. Several companies have either obtained permits, or are in the process of investing billions of dollars building facilities and obtaining permits that will allow them to export natural gas to any country in the world from the U.S. While plentiful supplies of natural gas in the U.S. keep the price around $3 per thousand cubic feet (mcf), other countries pay significantly more for their supply of natural gas. Europe pays around $11-12 per mcf and Japan is paying $16-17 per mcf. Most recently, the Commerce Department has authorized companies to export condensate – essentially an ultra-light crude oil with an API (American Petroleum Institute) gravity of 45 or higher.

In essence, the United States now exports most classes of hydrocarbon products - refined petroleum products, natural gas, manufactured products that use natural gas as a feedstock, and condensate (or ultra-light crude oil). Yet if the crude oil has an API gravity below 45, export is not allowed. This standard is arbitrary and depresses U.S. oil production. The restrictions force WTI (West Texas Intermediate) crude oil to be sold at a discount to Brent crude. Lifting the export ban on crude oil would boost U.S. production and would no longer unfairly penalize exploration and production companies, as well as mineral rights property owners in the U.S. - both of whom are currently forced to sell their crude oil only to U.S. and some Canadian customers.

Once again, I would like to thank the members of the committee for their kind attention, and the opportunity to speak before you today.

In: Mexico's Oil and Gas Industry
Editor: Neil Becker
ISBN: 978-1-63483-846-7

Chapter 4

TESTIMONY OF TONY PAYAN, DIRECTOR, MEXICO CENTER, JAMES A. BAKER III INSTITUTE FOR PUBLIC POLICY, RICE UNIVERSITY. HEARING ON "PURSUING NORTH AMERICAN ENERGY INDEPENDENCE: MEXICO'S ENERGY REFORMS"*

The purpose of my remarks today is to examine the basis for Mexico's recent energy reforms, to assess the major threats to its effective implementation and success, and to identify policy challenges that, if addressed, can strengthen the United States' relationship with Mexico.

ENERGY REFORM WAS NOT A CHOICE

Mexico's leaders did not deliberately choose to enact energy reforms. Rather, it was a policy shift forced by circumstances. Mexico has seen its oil production drop from about 3.5 million barrels per day (b/d) in 2008 to about 2.3 million b/d in 2015. At the same time, the country's energy consumption continues to grow. At current rates of production and consumption, Mexico

* This is an edited, reformatted and augmented version of testimony presented July 23, 2015 before the House Foreign Affairs Committee, Subcommittee on Western Hemisphere.

will become a net importer of oil somewhere around 2020.[1] Complicating matters, the country's federal budget is heavily dependent on oil revenues—about 30% to 35% of the federal budget comes from oil revenues.[2] Because Mexico's national oil company, PEMEX, handed over so much of its profits to the government, it could not reinvest in additional exploration and extraction of oil and gas, or in technology to tap into new sources of energy located in deep waters off the Gulf of Mexico. A no-change scenario would have put Mexico in a very difficult situation within a few years: it would have to spend valuable national resources to import oil and slash its federal budget by nearly one-third.

Thus, Mexico's energy reform, in my view, was primarily motivated by revenue maintenance and generation concerns, rather than a conviction that the free market would improve the country's energy future and advance economic development. Many of the contours of the reform demonstrate this intent. For example, the Mexican government, while reforming some aspects of PEMEX, continues to maintain tight control on the company. In the March 2015 Round Zero oil auction, PEMEX was the only bidder and received over 80% of the oil and gas plays it wished to keep.[3] What PEMEX did not want or could not keep was then fair game for others in subsequent bidding rounds. Second, the terms and conditions of contracts offered to foreign and private investors originally capped revenues, ensuring maximum revenue flows to Mexico's federal government. Companies pushed back. Since some of these caps on profits have been flexibilized. Third, the government has legislated that any contract can be rescinded at any point, even due to purely administrative violations—an undefined term, which will give federal bureaucrats enormous discretionary power over continuing, modifying or ending contracts. Fourth, the government intended to prohibit contract clauses that allow investors to bring contract disputes to international arbitration bodies, preferring to instead resolve disputes in Mexico's courts— which are notoriously deficient and susceptible to corruption. The courts' ability to fairly consider the interests of both investors and the Mexican government is questionable, to say the least. It is not clear yet whether Mexico's government has backed down on this point.

These examples are among many indicating that Mexico's government would have preferred to keep tight control of the energy sector rather than open it to foreign investors. However, circumstances forced the government to opt for reform. The reform, therefore, is a managed opening of the sector, rather than a full market-driven reform. It is likely that Mexico's government will try to maintain the upper hand in the energy sector, rather than allow market forces to determine production, consumption and profits. In my estimation, Mexico's energy reform is more restrictive than that of other countries, and continues to give the government excessive power over the sector. This should be kept closely in mind as Mexico's open energy market becomes an integral part of the North American energy landscape.

The results of Round One, Mexico's oil and gas play bidding process that took place July 15, 2015, were disappointing. Only two out of 14 blocks were sold. This is partly because the geopolitical market conditions of supply and demand are unfavorable to Mexico at this point; but it is also because there is much trepidation in the industry as to how to respond to the Mexican government's intent to retain the upper hand in the sector. This has introduced much uncertainty into the bidding process, and many industry actors have adopted a wait-and-see attitude.

Still, the situation in Mexico is not uncommon. Most countries heavily manage their energy sector, and IOCs have a long history of dealing with governments, even in difficult situations. Even so, Mexico presents unique challenges that must be examined closely. Although the country is not completely unpredictable, its institutions are not always accountable. As I attempt to break down these challenges, I will argue that, insofar as the U.S. is concerned, a much more comprehensive and forceful foreign policy toward our neighbor is required.

Major Threats to Energy Reform

Threats to energy reform in Mexico come from within the country itself. Its unique historical and structural problems might yet derail some of the positive effects of the historic reforms. In my view, three major problems threaten the success of energy reform implementation, and they must be considered as the U.S. examines its foreign policy toward Mexico: 1) Weakness of the rule of law and security issues; 2) Property rights, natural resources and social conflict; and 3) Corruption.

Weakness of the Rule of Law in Mexico and Security

The weakness of the rule of law in Mexico is historical and structural. These weaknesses must be understood in the context of the country's presidential system. During much of Mexico's existence, presidents have been more akin to strongmen than democratic rulers. During the 21st century,

Mexico's president was the central figure in the country's government and governance processes. The president dominated all of Mexico's institutions, including the central political party, Congress, state and local governments and the judiciary. The weakness and strength of Mexico's institutions were carefully calibrated to ensure that no one would challenge the central ruler during his six-year term, and that the powerful PRI party would stay in office. In 2000, when Mexico made the leap to a more democratic transition by voting into office an opposition party, most institutions were able to reconstitute themselves; they function relatively well and independently of the weakened executive branch, except the judiciary. The police and judiciary systems in Mexico remain structurally weak. They are ineffective, underfunded and highly susceptible to corruption.

The weakness of the civilian police-justice system in Mexico has emboldened organized crime in the country. The last 10 years have seen an upsurge in criminal violence that has spread to vast swaths of the Mexican territory. Criminals, as we have seen with the recent escape of Joaquín "El Chapo" Guzmán Loera, feel they can corrupt nearly any official and strike at the State almost anywhere at will.

U.S.-Mexico agreements, such as the Mérida Initiative, which have focused primarily on drug trafficking, have done little to overcome the structural weaknesses of the civilian police and justice systems in Mexico. Instead, the effect has been to break up Mexico's organized crime organizations into dozens of smaller, powerful groups that control cities and corridors and that have diversified their criminal activities to include kidnapping for ransom, extortion, human smuggling and trafficking and, of course oil, gas and fuel theft.

This is, in fact, one of the greatest sources of concern for foreign and private investors seeking to enter Mexico's newly opened energy sector: they fear their investments, workforces, equipment and products may eventually be tapped into by organized crime. There is much unease with this situation, and it may have contributed to some companies taking their time to examine Mexico more carefully before going into business there.

In hindsight, U.S. efforts to prod Mexico into fighting hierarchical criminal organizations have also contributed to the fragmentation of these organizations into dozens of smaller criminal gangs and to the diversification of their criminal activities. Such a result can be viewed as a way to eventually send the military back to the barracks and allow local police forces to face down the smaller gangs of criminals. However, this has not happened. The local police forces, some 400,000 officers in the country, are underpaid, underequipped, unprepared, probably undervalued and generally incapable of confronting the well-organized, disciplined and well-armed gangs that dominate cities, towns and corridors throughout the country. This situation makes the police particularly susceptible to corruption. The problem of criminal groups in Mexico is, as a consequence, much more difficult to control.

Property Rights, Natural Resources and Social Conflict

In Mexico, individuals own the use of the land, but do not own the rights to the water or oil beneath it. Two major issues may emerge out of this. One is the fact that foreign companies will have to negotiate with landowners to access natural resources. Considering that 51% of Mexico's land ownership is social—either in the form of *ejidos*[4] or as indigenous lands—this can also mean that many companies will be forced to negotiate not with individual owners, but with entire farmers' groups or indigenous groups over issues such as right-of-way, water availability, use of the land, reparation and remediation. This is a potentially explosive situation that may give rise to social conflict, particularly as landowners resist granting a right-of-way or permanently or temporarily ceding their land for energy projects, as the law mandates. Worse, the law gives landowners 180 days to agree to grant access to minerals under their land, or the State will force them to accept a third-party negotiator. This can be a recipe for social conflict.

There seems to be another conflict in Mexican law that needs further investigation. Whereas energy reform gives energy-related projects priority over the use of natural resources under the theory that it is in the interest of national development, the water law in Mexico states that access to water is a human right and human access to water is a priority over other uses. This is not likely to affect the large oil corporations—many of which will invest in deep waters in the Gulf of Mexico in any event—but it may become relevant once oil and gas prices recover and the shale in northern Mexico, where water is extremely scarce, begins to be developed. In this regard, it is possible that social conflicts over water resources may develop in the future, particularly in northern Mexico, right across the Texas border.

Corruption

So far, the Mexican National Hydrocarbon Commission and other energy authorities in Mexico have done an excellent job of maintaining the process of opening the energy sector in a way that is both transparent and responsive to the concerns of the industry. However, corruption by the political class is another concern in the implementation of energy reform. In fact, President Peña has himself been involved in several corruption scandals that have undermined the legitimacy of the energy reform as well as of other reforms. As of yet, there is no effective anti-corruption system in Mexico capable of dealing with governors and mayors, many of whom are notoriously corrupt and will ultimately have the ability to make the life of investors in the field very easy or very difficult. Many of them will also be seeking opportunities to use their political power to maneuver themselves into deals as partners with private and foreign investors.

The failure of Round One this month cannot be attributed to a lack of effort on the part of the Mexican federal government to be clear about the terms of business and to respond to the concerns of industry. But this transparency and responsiveness cannot be guaranteed in the long term, or even in the medium term, in a system that suffers from endemic corruption, and where there are hundreds of actors at the federal, state and local level with the ability to affect energy project development throughout the country. Corruption in Mexico has become a lifestyle. Mayors, governors, members of Congress and politicians of all sorts tend to view public service as a way to serve themselves as well as the Mexican people. They have embezzled public funds; used their political power to ensure privileges for themselves, their families and friends; and inappropriately used their public position to engage business interests to enrich themselves. By almost any measure, Mexico ranks among the most corrupt countries in the world.[5] This is one of the most important concerns of the energy industry—that they may be liable under U.S. law for corrupt practices in Mexico. Their concern is justified, as they will eventually have to deal with hundreds of politicians at all levels. Although the energy sector is under federal jurisdiction, the energy projects themselves have a fundamentally local character. They take place in communities across the country under the eye of citizens and politicians everywhere.

This does not mean that the U.S. should let up on enforcing its foreign corrupt practices laws, but rather, that it should become more attuned to the challenges that the industry will face in a country where corruption is wide and tolerated.

CONCLUSION

To conclude, it is important to note that the failure of Round One this month is an inauspicious sign. Energy reform will move forward, but it will experience more difficulties than expected. The structure of incentives built into contracts, the natural resource conditions, property rights, and the weak rule of law all threaten the smooth implementation of energy reform in Mexico.

Without a comprehensive foreign policy toward Mexico in the energy arena, there is very little the United States will be able to do to help this situation. Until now, Mexico has been dealt with very tentatively, given the country's nationalistic sensibilities. The Peña administration in particular is even more sensitive to American activities in Mexico than the two previous PAN administrations were. But there is hardly any country that has more leverage on Mexico than the United States. Thus, the U.S. government should now reexamine its relationship with Mexico and begin to put together a policy that goes beyond trade, economics and the drug war. Mexico and the U.S. must now engage in policies that emphasize institutional development, democracy-building measures and an intense campaign to fight corruption, as well as binational human and labor mobility and the creation of binational institutions that work for both nations. Below I outline some basic policy recommendations.

Policy Recommendations

1) The Mérida Initiative should be replaced with a more comprehensive policy toward Mexico.

   The Mérida Initiative has not paid off. Although there are some isolated victories, it is now clear that the Mérida Initiative has not had its intended effects. While it has helped break up most of the large criminal organizations that existed in Mexico, it has created a more fragmented and chaotic criminal landscape. There are now anywhere between 60 and 80 criminal organizations, some dedicated to drug trafficking, notably the Sinaloa Cartel, but dozens of them are now dedicated to victimizing Mexican citizens through kidnapping, extortion, robberies and theft. Some of these organizations have special units dedicated to oil, gas and fuel theft. We have also seen some unscrupulous U.S. business owners buy some of that ill-gotten oil from Mexican criminals, notably in Texas. Energy investors, and particularly those that will invest on onshore projects, will necessarily confront some of these organizations. Their equipment, their personnel, and in general their investment will come under fire here and there.

2) Focus American foreign policy on strengthening Mexico's institutions, particularly the police and judicial systems.

   Mexico's government has made very little progress in implementing the 2008 Judicial Reform.[6] Although some states have made substantial progress, most of them lag well behind, so the Judicial Reform will likely not be implemented by 2016, as mandated by the law. The United States must now urge Mexico not only to implement Judicial Reform but also to implement a more forceful anti-corruption program that primarily focuses on making the judiciary a truly independent branch of government, with the capability, the resources and the political will to investigate and punish corruption wherever it may be found. Without a strong and independent judiciary capable of investigating political and other corruption, Mexico will not be a reliable partner in achieving a North American energy plan or a broader North American economic bloc.

3) Aggressively include Mexico's political class in corruption investigations.

   The United States is the destination of many in Mexico's corrupt political class. They freely move about in the United States; they do business in the United States; they launder their ill-gotten profits in the United States; they send their children to the United States; and they generally use the United States to seek refuge when they face complications in Mexico. Policy toward Mexico must now include a serious effort to address corruption in Mexico by investigating the country's political class and targeting them with anti-corruption efforts. There should be no access to the United States to anyone accused of malfeasance in Mexico.

4) The U.S. must develop a more comprehensive policy toward Mexico, one that considers labor integration, manufacturing chains, energy swaps in oil and gas, and financial sector integration, along with additional funding targeted to the promotion of institutional reform in Mexico's police and judicial system. Trade between the two countries is a story of success, but we have not been able to use this success to ensure that its dividends quickly spread to other areas, and we have not used that momentum to create truly regional institutions that will help us achieve a fully developed and democratic North America, one that will work for all Mexicans and Americans. Mexico's energy opening is yet another important step, which we must take advantage of to make a final push to create the most prosperous region in the world.

End Notes

[1] Kenneth B. Medlock III and Donald Soligo, "The Future of Oil in Mexico: Scenario for Oil Supply, Demand and Net Exports for Mexico," James A. Baker III Institute for Public Policy at Rice University, Houston, Texas, April 29, 2015.

[2] Mexican Republic Decree on the Federal Budget for FY2015. November 13, 2014, http://www.hacienda.gob.mx/INGRESOS/Ingresos_ley/2015/lif_2015.pdf.

[3] See Secretaría de Energía, "Results of Technical Analysis," found at http://www.energia.gob.mx/webSener/rondacero_Ingles/9600.html.

[4] Ejidos are communally owned lands, assigned by the government to farmers and ranchers to manage collectively during the agrarian reform laws of the 20th century. Although legislation in the early 1990s allowed many to sell their land, numerous ejidos are still operational.

[5] Mexico is ranked 103/175 by Transparency International, with a score of 35/100. There is little enforcement of anti-corruption laws. See http://www.transparency.org/country#MEX.

[6] Trans-Border Institute. Joan B. Kroc School of Peace Studies. University of San Diego. Judicial Reform in Mexico: Change and Challenges in the Justice Sector. 2010.

In: Mexico's Oil and Gas Industry
Editor: Neil Becker
ISBN: 978-1-63483-846-7

Chapter 5

TESTIMONY OF ERIC FARNSWORTH, VICE PRESIDENT, COUNCIL OF THE AMERICAS. HEARING ON "PURSUING NORTH AMERICAN ENERGY INDEPENDENCE: MEXICO'S ENERGY REFORMS"*

Good afternoon, Mr. Chairman, Mr. Ranking Member, and Members. It is a privilege to appear before you again. I am pleased to offer testimony on North American energy independence and the importance of Mexico's ongoing reform efforts. Thank you for the opportunity to join this timely and important hearing.

As I suggested during oversight testimony in February before this Subcommittee, our most immediate neighbors have the greatest impact on our strategic interests. It is significant and consequential that we have a northern neighbor, Canada, which is both our top trade partner and top energy supplier, and which is stalwart in working with us to address the toughest global issues. Our southern neighbor, Mexico, is our second largest export market and also third largest energy supplier. Together, the three nations of North America are quickly becoming a fully integrated production platform, with tightly-knit supply chains and ever-deepening commercial ties. As China and others have emerged to challenge the pre-existing global economic order, the ties that bind

* This is an edited, reformatted and augmented version of testimony presented July 23, 2015 before the House Foreign Affairs Committee, Subcommittee on Western Hemisphere.

us within North America have created economic efficiencies, built competitiveness, and directly contributed to the economic growth that our citizens both desire and deserve.

Access to affordable, stable energy supplies has been a fundamental driver of strategic U.S. policy on a bipartisan basis at least since the 1930's. Actions that the three North American governments take to increase regional energy integration contribute meaningfully to support and expand our competitiveness in a manner consistent with U.S. interests. Mexico's decision to open its energy markets and link them more closely to its North American neighbors is therefore a historic and meaningful action for Mexico itself as well as a strategically important step for the entire community of North America.

Mexico's Energy Reforms are Transformational

It would be difficult to overestimate the importance of Mexico's potentially transformational energy reforms, once fully implemented. The Mexican energy sector is historically quite sensitive politically. It was closed to foreign investors for over 75 years. The sector would benefit from additional technology, management expertise, and capital to address production declines and high costs, which are a drag on competitiveness and economic performance.

Since 2013, the Peña Nieto administration has been working to address this, passing and now beginning to implement legislation allowing for private participation in both the oil and gas and also power generation sectors. At the same time, the government has expressed awareness of the need for full transparency in the reform and bidding process in order to build broad public faith that the final outcome will be fair and free from corruption and sweetheart dealing. This is a critical point, upon which the sustainability of the entire reform effort ultimately depends.

Since the beginning of the reform effort, the price of crude has fallen approximately 50 percent, which impacts investment decisions particularly in sectors as capital intensive as energy. Investors become more selective and the quality of investment opportunities becomes paramount. There have also been recent setbacks regarding the rule of law in Mexico, notably the mid-July escape from prison of Joaquín "El Chapo" Guzmán. While not directly related to the energy sector, this does play into a narrative concerning the risks of doing business in Mexico with a particular focus on corruption and the rule of law.

Taken together, these data points have led some observers to turn more bearish on Mexico's energy prospects. Some of these issues may have been in play on July 15 for Mexico's first round of investor bidding, in which only two of 14 blocks offered to investors were awarded. But I would submit that this focus misses the bigger picture.

INVESTMENT DECISIONS FOR THE LONGER RUN

The first bidding round featured marginal shallow water blocks, with prices arbitrarily set by the government. Future auctions will feature more attractive deep water and onshore unconventional shale oil and gas blocks. Lessons learned from the first auction on pricing and other matters will also be applied to later rounds. As well, on a comparative basis globally, additional investment blocks are larger and potentially more significant than those available elsewhere in the world.

Investor interest in the deep water reservoirs of the Gulf of Mexico and unconventional shale oil and gas in the Eagle Ford shale in the Burgos Basin is high. Mexico's shale reserves are the sixth largest in the world, according to the U.S. Energy Information Administration, overlapping with shale resources in southern Texas. Just as the shale revolution increased U.S. oil production by more than 55 percent, the application of new technology, capital, and best management practices could catalyze a similar jump in oil and gas production in Mexico.

Temporary set-backs aside, Mexico's energy reform efforts are a broader play which will continue to attract interest.

POWER GENERATION AND THE ENVIRONMENT

While most attention has been on the oil and gas sector, power generation is also being liberalized, and this may prove to be an even bigger story because the cost of electricity in Mexico is well above global rates. This raises overall production costs making Mexico less attractive in manufacturing and other activities, while also reducing disposable income available to consumers who might otherwise reallocate their spending. While previously only the Federal

Electricity Commission (CFE) was allowed to generate electricity, the sector is now opening to private investors. Overall, energy infrastructure must be upgraded, including pipelines, rail, waterways, and transmission lines, among others, and CFE recently announced that it would tender projects for close to $10 billion. As investment increases and prices go down, Mexico will become more competitive as resources are directed toward more efficient and productive uses.

Power generation reforms also directly impact broader issues, in particular the environment and global climate change. Updating infrastructure allows for more efficient power generation, utilizing cleaner and more renewable fuel inputs. Of course, conservation is the cleanest and most cost-effective fuel available, and should be actively promoted in addition to efficiencies. Beyond this, Mexico's reforms promote renewable energy development, and investors have shown a keen interest in accessing this market, especially wind, solar, and geothermal. Regulations are sound and provide a clear framework for investment so long as targets are flexible. As well, clean energy targets can contribute to overall energy targets, both on a North American and a sub-regional basis.

MEXICO'S REFORMS POSITIVELY IMPACT NORTH AMERICA

The success of Mexico's energy reforms is important for the United States and for North America broadly. The North American production platform is already integrating content produced across borders; some 40 percent of the content of goods exported to the United States from Mexico is originally U.S. content. From Canada, the figure is 25 percent. From China, it is merely four percent.

Actions which contribute to Mexican competitiveness can therefore help to increase our own economic well-being. In particular, investment restrictions to date mean that natural gas in Mexico is more expensive than in the United States, increasing production costs in sectors that use gas as a feedstock such as chemicals as well as in all sectors that draw from the power grid. This reduces manufacturing competitiveness. Gas imports from the United States have helped, although inbound pipelines are functioning at capacity. More could be constructed, but a better solution would be to develop Mexico's own energy resources, which reforms are intended to do. Substituting cleaner fuels such as gas for diesel would also support national and regional environmental targets. Once fully implemented, prices would fall and the North American manufacturing platform would become more competitive vis-à-vis global production centers such as China and Europe.

The U.S. Role

In support of Mexico's reforms, the United States has an important role to play. In the first instance, Congress could strengthen U.S.-Mexico energy relations by lifting crude oil export restrictions toward a proposed swap between Mexico and the United States, a procedure already allowed with Canada. Mexico would supply heavy crude in exchange for U.S. light sweet crude to best take advantage of refining capacity. Not only would the swap be a welcome move toward dropping the U.S. crude oil export ban, but, symbolically, would provide Mexico similar treatment as our other NAFTA partner.

More broadly, regional energy integration must be a priority not an afterthought, underlining an intentional drive for economic expansion. Within this construct, regulatory alignment is critical, infrastructure must be upgraded, investment climate issues including an emphasis on workforce development should be addressed, and joint research and development projects expeditiously pursued.

Finally, in the run up to the Paris meetings in December, a regional approach to climate change issues would be useful and should be pursued as a means to engage the global dialogue. In March, the United States and Mexico released a joint statement on climate cooperation, launching a clean energy and climate task force. In May, North America's energy ministers established a working group seeking to harmonize relevant regulations.

These are solid steps. More can be done. The key to this overall agenda is the successful implementation of Mexico's liberalizing reforms. It is manifestly in our own interests that the Mexican people succeed. We should assist the process where appropriate and wherever we can.

I look forward to your questions.

In: Mexico's Oil and Gas Industry
Editor: Neil Becker
ISBN: 978-1-63483-846-7

Chapter 6

U.S.-MEXICO ECONOMIC RELATIONS: TRENDS, ISSUES, AND IMPLICATIONS*

M. Angeles Villarreal

SUMMARY

During the 114th Congress, policy makers will likely maintain an interest in Mexico on issues related to cross-border trade and investment; Mexico's economic reform measures, especially in the energy sector; the Trans-Pacific Partnership (TPP) agreement negotiations; NAFTA and WTO trade issues; and U.S.-Mexico border management. Congress may also take an active interest in ongoing bilateral efforts to promote economic competitiveness, increase regulatory cooperation, and pursue energy integration. Under the U.S.-Mexico High Level Economic Dialogue (HLED), which was launched in September 2013, the United States and Mexico are striving to advance economic and commercial priorities through annual meetings at the Cabinet level that also include leaders from the public and private sectors. Another bilateral initiative that may be of interest to policy makers is the High-Level Regulatory Cooperation Council (HLRCC), launched in February 2012, which is intended to help align regulatory principles. In addition, the two countries have a bilateral initiative for improving border management under the

---

* This is an edited, reformatted and augmented version of a Congressional Research Service publication RL32934, prepared for Members and Committees of Congress, dated April 20, 2015.

Declaration Concerning 21st Center Border Management that was announced in 2010.

The economic and trade relationship with Mexico is of interest to U.S. policy makers because of Mexico's proximity to the United States, the high level of bilateral trade, and the strong cultural and economic ties that connect the two countries. Also, it is of national interest for the United States to have a prosperous and democratic Mexico as a neighboring country. Mexico is the United States' third-largest trading partner, while the United States is, by far, Mexico's largest trading partner. Mexico ranks third as a source of U.S. imports, after China and Canada, and second, after Canada, as an export market for U.S. goods and services. The United States is the largest source of foreign direct investment (FDI) in Mexico.

The United States and Mexico have strong economic ties through the North American Free Trade Agreement (NAFTA), which has been in effect since 1994. Most studies show that the net economic effects of NAFTA on both countries have been small but positive, though there have been adjustment costs to some sectors within both countries. Much of the bilateral trade between the United States and Mexico occurs in the context of production sharing as manufacturers in each country work together to create goods. The expansion of trade has resulted in the creation of vertical supply relationships, especially along the U.S.-Mexico border. The flow of intermediate inputs produced in the United States and exported to Mexico and the return flow of finished products greatly increased the importance of the U.S.-Mexico border region as a production site. U.S. manufacturing industries, including automotive, electronics, appliances, and machinery, all rely on the assistance of Mexican manufacturers.

In June 2012, Mexico and Canada joined negotiations for the proposed TPP. If approved, the agreement would likely enhance the economic links Mexico already has with the United States and Canada under NAFTA. Policy makers may consider how a TPP would affect the U.S.-Mexico trade and investment relationship under NAFTA. Although nearly all U.S. trade with Mexico is now conducted duty and barrier free under NAFTA, the TPP negotiations may provide a venue for addressing issues that are not covered under the agreement. If approved, a TPP would not render NAFTA obsolete, but it could alter certain rules governing U.S.-Mexico trade since NAFTA entered into force. A TPP may have implications in several areas, including intellectual property rights protection, investment, state-owned enterprises, services trade, agriculture, government procurement, worker rights, and the environment.

INTRODUCTION

The bilateral economic relationship with Mexico is of key interest to the United States because of Mexico's proximity, the high volume of trade with Mexico, and the strong cultural and economic ties between the two countries. The United States and Mexico share many common interests related to trade, investment, and regulatory cooperation. The two countries share a 2,000 mile border and have extensive interconnections through the Gulf of Mexico. There are also links through migration, tourism, environmental issues, health concerns, and family and cultural relationships.

The 114th Congress will likely maintain an active interest in Mexico on issues related to cross-border trade and investment; Mexico's economic reform measures, especially in the energy sector; the Trans-Pacific Partnership (TPP) agreement negotiations; NAFTA and WTO trade issues; and U.S.-Mexico border management. Congress may continue to take an interest in the economic policies of Mexico's President, Enrique Peña Nieto. Since entering into office on December 1, 2012, Peña Nieto has advocated numerous economic and political reforms that include, among other measures, opening up the energy sector to private investment, countering monopolistic practices, passing fiscal reform, making farmers more productive, and increasing infrastructure investment.[1] Peña Nieto also endorses an active international trade policy aimed at increasing Mexico's trade with Asia, South America, and other markets. His government is taking an active role in the negotiations for a TPP.[2]

This report provides an overview of U.S.-Mexico economic relations, trade trends, the Mexican economy, NAFTA, and trade issues between the United States and Mexico.

U.S.-MEXICO ECONOMIC RELATIONS

Mexico is one of the United States' key trading partners, ranking second among U.S. export markets and third in total U.S. trade (imports plus exports). Under the North American Free Trade Agreement (NAFTA), the United States and Mexico have developed significant economic ties. Trade between the two countries more than tripled since the agreement was implemented in 1994. Through NAFTA, the United States, Mexico, and Canada form one of the world's largest free trade areas, with about one-third of the world's total gross domestic product (GDP). Mexico has the second-largest economy in Latin America after Brazil. It has a population of 124 million people, making it the most populous Spanish-speaking country in the world and the third-most populous country in the Western Hemisphere (after the United States and Brazil).

Mexico's gross domestic product (GDP) was an estimated $1.3 trillion in 2014, about 7% of U.S. GDP of $17.4 trillion. In purchasing power parity[3], Mexico's GDP is higher, $2.1 trillion in 2014 or about 12% of U.S. GDP. Per capita income in Mexico is significantly lower than in the United States. In 2014, Mexico's per capita GDP in purchasing power parity was $16,773, or 69% lower than U.S. per capita GDP of $54,634 (see *Table 1*). Ten years earlier, in 2004, Mexico's per capita GDP in purchasing power parity was $11,080, or 74% lower than the U.S. amount of $41,887. Although there is a notable income disparity with the United States, Mexico's per capita GDP is relatively high by global standards, and falls within the World Bank's upper-middle income category.[4] Mexico's economy relies heavily on the United States as an export market. The value of exports equaled 33% of Mexico's GDP in 2014, as shown in *Table 1*, and approximately 80% of Mexico's exports are headed to the United States.

U.S.-Mexico Trade

The United States is, by far, Mexico's leading partner in merchandise trade, while Mexico is the United States' third-largest trade partner after China and Canada. Mexico ranks second among U.S. export markets after Canada, and is the third-leading supplier of U.S. imports. U.S. trade with Mexico increased rapidly since NAFTA entered into force in January 1994. U.S. exports to Mexico increased from $41.6 billion in 1993 (the year prior to NAFTA's entry into force) to $240.3 billion in 2014, an increase of 478%. Imports from Mexico increased from $39.9 billion in 1993 to $294.2 billion in 2014, an increase of 637% (see *Figure 1*). The merchandise trade balance with Mexico went from a surplus of $1.7 billion in 1993 to a widening deficit that reached a peak of $74.3 billion in 2007. In 2014, the merchandise trade deficit with Mexico was $53.8 billion. In services, the United States had a surplus of $12.1 billion in 2013. U.S. exports in services to Mexico totaled $29.9 billion in 2013, while U.S. imports totaled $17.8 billion.[5]

Table 1. Key Economic Indicators for Mexico and the United States

	Mexico		*United States*	
	2004	*2014*[a]	*2004*	*2014*
Population (millions)	109	124	293	319
Nominal GDP (US$ billions)[b]	770	1,283	12,275	17,419
Nominal GDP, PPPc Basis (US$ billions)	1,212	2,077	12,275	17,419
Per Capita GDP (US$)	7,040	10,360	41,887	54,634
Per Capita GDP in $PPPs	11,080	16,773	41,887	54,634
Nominal exports of goods & services (US$ billions)	202	420	1,182	2,337
Exports of goods & services as % of GDP[d]	26%	33%	10%	13%
Nominal imports of goods & services (US$ billions)	216	432	1,801	2,875
Imports of goods & services as % of GDP[d]	28%	34%	15%	17%

Source: Compiled by CRS based on data from Economist Intelligence Unit (EIU) online database.

[a] Some figures for 2014 are estimates.

[b] Nominal GDP is calculated by EIU based on figures from World Bank and World Development Indicators.

[c] PPP refers to purchasing power parity, which reflects the purchasing power of foreign currencies in U.S. dollars.

[d] Exports and Imports as % of GDP derived by EIU.

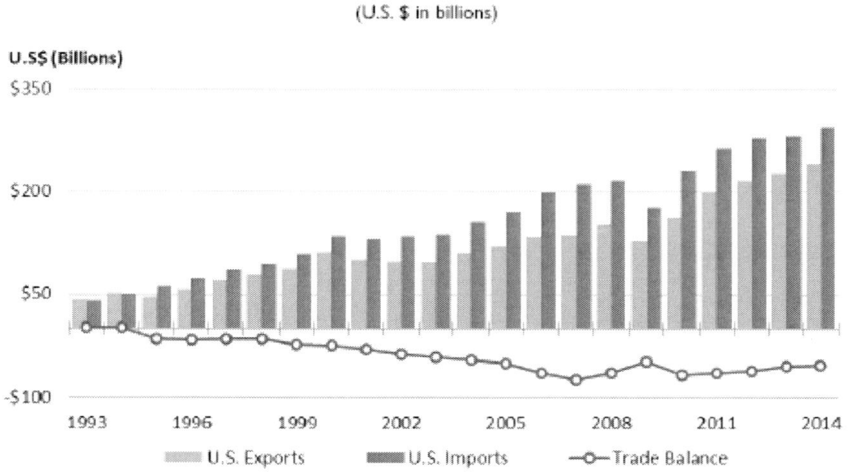

Source: Compiled by CRS using the United States International Trade Commission (USITC) Interactive Tariff and Trade DataWeb at http://dataweb.usitc.gov.

Figure 1. U.S. Merchandise Trade with Mexico.

Mexico's reliance on the United States as an export market has diminished very slightly over the years with the percentage of Mexico's total exports in goods and services going to the United States decreasing from 84% in 1996 to 79% in 2013. Mexico's share of the U.S. market has lost ground since 2003 when China surpassed Mexico as the second-leading supplier of U.S. imports. The U.S. share of Mexico's import market has also decreased. Between 1996 and 2013, the U.S. share of Mexico's total imports decreased from 83% to 54%. China is Mexico's second-leading supplier of imports.[6]

The leading U.S. import item from Mexico in 2014 was motor vehicles ($46.4 billion), followed by motor vehicle parts ($40.1 billion), oil and gas ($27.8 billion), computer equipment ($14.3 billion), and audio and video equipment ($14.2 billion), as shown in Table 2. After sharp decreases in 2009 caused by the global economic downturn, U.S. imports from Mexico increased from $176.5 billion in 2009 to $294.2 billion in 2014. The leading U.S. export item to Mexico in 2014 was motor vehicle parts ($21.5 billion), followed by petroleum and coal products ($19.1 billion), computer equipment ($16.0 billion), semiconductors and other electronic components ($13.5 billion), and basic chemicals ($10.1 billion), as shown in Table 3.

Table 2. U.S. Imports from Mexico: 2009-2013

(U.S. $ in billions)

Items (NAIC 4-digit)	*2010*	*2011*	*2012*	*2013*	*2014*	*% Total in 2014*
Motor vehicles	27.5	30.5	35.3	40.1	46.4	16%
Motor vehicle parts	23.6	28.5	33.3	36.2	40.1	14%
Oil and gas	29.7	39.6	37.2	32.0	27.8	9%
Computer equipment	13.6	14.5	16.0	14.8	14.3	5%
Audio and video equipment	16.5	14.7	14.2	13.8	14.2	5%
Other	119.1	135.0	141.5	143.6	151.4	51%
Total	230.0	262.9	277.6	280.5	294.2	

Source: Compiled by CRS using USITC Interactive Tariff and Trade DataWeb at http://dataweb.usitc.gov: North American Industrial Classification (NAIC) 4-digit level.

Note: Nominal U.S. dollars.

Table 3. U.S. Exports to Mexico: 2010-2014
(U.S. $ in Billions)

Items (NAIC 4-digit)	*2010*	*2011*	*2012*	*2013*	*2014*	*% Total in 2014*
Motor vehicle parts	14.0	17.2	19.4	21.1	21.5	9%
Petroleum and coal products	12.0	20.3	20.7	19.3	19.1	8%
Computer equipment	9.9	13.4	14.5	14.8	16.0	7%
Semiconductors and other electronic components	11.8	10.9	11.4	13.0	13.5	6%
Basic chemicals	7.1	9.1	10.1	10.1	10.1	4%
Other	108.8	127.4	139.9	147.7	160.2	67%
Total	163.7	198.3	215.9	226.1	240.3	

Source: Compiled by CRS using USITC Interactive Tariff and Trade DataWeb at http://dataweb.usitc.gov: NAIC 4-digit level.

Note: Nominal U.S. dollars.

Table 4. U.S.-Mexican Foreign Direct Investment Positions: 1994-2013 Historical Cost Basis
(U.S. $ in millions)

Year	Mexican FDI in the U.S.	U.S. FDI in Mexico
1994	2,069	16,968
1995	1,850	16,873
1996	1,641	19,351
1997	3,100	24,050
1998	2,055	26,657
1999	1,999	37,151
2000	7,462	39,352
2001	6,645	52,544
2002	7,829	56,303
2003	9,022	56,851
2004	7,592	63,384
2005	3,595	73,687
2006	5,310	82,965
2007	8,478	91,046
2008	8,420	87,443
2009	11,111	84,047
2010	10,970	85,751
2011	12,500	85,599

Table 4. (Continued)

Year	Mexican FDI in the U.S.	U.S. FDI in Mexico
2012	14,458	98,377
2013	17,610	101,454

Source: U.S. Department of Commerce, Bureau of Economic Analysis.

The sharp rise in U.S. investment in Mexico since NAFTA is also a result of the liberalization of Mexico's restrictions on foreign investment in the late 1980s and the early 1990s. Prior to the mid-1980s, Mexico had a very protective policy that restricted foreign investment and controlled the exchange rate to encourage domestic growth, affecting the entire industrial sector. Mexico's trade liberalization measures and economic reform in the late 1980s represented a sharp shift in policy and helped bring in a steady increase of FDI flows into Mexico. NAFTA provisions on foreign investment helped to lock in the reforms and increase investor confidence. Under NAFTA, Mexico gave U.S. and Canadian investors nondiscriminatory treatment of their investments as well as investor protection. NAFTA may have encouraged U.S. FDI in Mexico by increasing investor confidence, but much of the growth may have occurred anyway because Mexico likely would have continued to liberalize its foreign investment laws with or without the agreement.

Mexico's Export-Oriented Assembly Plants

Mexico's export-oriented assembly plants are closely linked to U.S.-Mexico trade in various labor-intensive industries such as auto parts and electronic goods. These plants generate a large amount of trade with the United States, and a majority of the plants have U.S. parent companies. Foreign-owned assembly plants, which originated under Mexico's maquiladora program in the 1960s,[7] account for a substantial share of Mexico's trade with the United States. The border region with the United States has the highest concentration of assembly plants and workers. Most maquiladoras currently export the majority of their production to the U.S. market. Prior to NAFTA, a maquiladora was limited to selling up to 50% of the previous year's export production to the domestic market.

Private industry groups state that these operations help U.S. companies remain competitive in the world marketplace by producing goods at competitive prices. In addition, the proximity of Mexico to the United States allows production to have a high degree of U.S. content in the final product, which could help sustain jobs in the United States. Critics of these types of operations argue that they have a negative effect on the economy because they take jobs from the United States and help depress the wages of low-skilled U.S. workers.

NAFTA, along with a combination of other factors, contributed to maquiladora growth after 1993. Trade liberalization, wages, and economic conditions, both in the United States and Mexico, increased the number of Mexican export-oriented assembly plants. Although some provisions in NAFTA may have encouraged growth in certain sectors, manufacturing activity likely has been more influenced by the strength of the U.S. economy and relative wages in Mexico.

Changes in Mexican regulations on export-oriented industries after NAFTA merged the maquiladora industry and Mexican domestic assembly-for-export plants into one program called the Maquiladora Manufacturing Industry and Export Services (IMMEX). In 2001, the North American rules of origin determined the duty-free status for a given import and replaced the previous special tariff provisions that applied only to maquiladora operations. The initial maquiladora program ceased to exist and the same trade rules applied to all assembly operations in Mexico.

NAFTA rules for the maquiladora industry were implemented in two phases, with the first phase covering the period 1994-2000, and the second phase starting in 2001. During the initial phase, NAFTA regulations continued to allow the maquiladora industry to import products duty-free into Mexico, regardless of the country of origin of the products. This phase also allowed maquiladora operations to increase maquiladora sales into the domestic market. Phase II made a significant change to the industry in that the new North American rules of origin determined duty-free status for U.S. and Canadian products exported to Mexico for maquiladoras. The elimination of duty-free imports by maquiladoras from non-NAFTA countries under NAFTA caused some initial uncertainty for the companies with maquiladora operations. Maquiladoras that were importing from third countries, such as Japan or China, would have to pay applicable tariffs on those goods under the new rules.

Worker Remittances to Mexico

Remittances are one of the three highest sources of foreign currency for Mexico, along with oil and tourism. Most remittances to Mexico come from workers in the United States who send money back to their relatives in Mexico. Mexico receives the largest amount of remittances in Latin America. Remittances are often a stable financial flow for some regions in Mexico as workers in the United States make efforts to send money to family members. Most of the remittances going to Mexico go to southern states in Mexico where poverty levels are high. Studies indicate that women are the primary recipients of the money, and usually use it for basic needs such as rent, food, medicine, or utilities.[8]

Annual remittances to Mexico decreased from $22.4 billion in 2012 to $21.7 billion in 2013, as shown in *Table 5*.[9] In 2009 remittances experienced a sharp decline of 15.2%, likely due to the global financial crisis. Prior to this, remittances to Mexico had been increasing rapidly. Between 1996 and 2007, remittances increased from $4.2 billion to $25.1 billion, an increase of over 500%. The annual growth rate reached a high of 54.3% in 2003, and then continued at a slower rate until 2008, when the rate of growth declined. The drop in remittances could be related to changes in migration flows as well to increases in the exchange rate between the Mexican peso and the U.S. dollar.[10]

Table 5. Percent Changes in Remittances to Mexico

(U.S. $ in billions)

	2002	2003	2004	2005	2006	2007	2008	2009	2010	2011	2012	2013
Amount	9.8	15.1	18.3	21.7	25.6	26.1	25.1	21.3	21.3	22.8	22.4	21.7
% Change	10.2%	54.3%	21.1%	18.3%	17.9%	1.9%	-3.5%	-15.2%	0.0%	7.0%	-1.8%	-3.1%

Source: Compiled by CRS using data from the Inter-American Development Bank, Multilateral Investment Fund; and Mexico's Central Bank.

Electronic transfers and money orders are the most popular methods to send money to Mexico. The rapid increase in remittances during the late 1990s through the mid-2000s can be attributed to numerous factors, but it was also largely influenced by considerable reductions in transaction fees charged by banks. In the 1990s, these fees were as high as 8%, and went down as low as 2.5% in 2003.[11] The Inter-American Development Bank reported that the average cost to send $200 was 6.0% in 2010.[12]

Worker remittance flows to Mexico have an important impact on the Mexican economy, in some regions more than others. Some studies on remittance flows to Mexico report that in southern Mexican states, remittances mostly or completely cover general consumption and/or housing. A significant portion of the money received by households goes for food, clothing, health care, and other household expenses. Some remittances mat be used for capital invested in microenterprises throughout urban Mexico. The economic impact of remittance flows is concentrated in the poorer states of Mexico.

Bilateral Economic Cooperation

The Obama Administration has engaged in bilateral efforts with Mexico, and also with Canada, to address issues related to the U.S.-Mexico border, enhance economic competitiveness, increase regulatory cooperation, and pursue energy integration.

High Level Economic Dialogue (HLED)

On September 20, 2013, the United States and Mexico launched the High Economic Dialogue (HLED), co-chaired by the U.S. Department of State, Department of Commerce, the Office of the United States Trade Representative, and their Mexican counterparts. [13] The purpose of the initiative is to advance U.S.-Mexico economic and commercial priorities that are central to promoting mutual economic growth, job creation, and global competitiveness. The HLED is organized around three broad pillars, including: 1) promoting competitiveness and connectivity; 2) fostering economic growth, productivity, and innovation; and 3) partnering for regional and global leadership. The HLED's major goals are to promote competitiveness in specific sectors such as transportation, telecommunications, and energy; to explore ways to promote entrepreneurship and encourage the development of human capital; and to facilitate greater alignment on issues of shared concern in both regional and international initiatives, especially in trade negotiation. Some of the initial steps toward accomplishing such goals have included: continuing the work of the Mexico-U.S. Entrepreneurship and Innovation Council; collaborating on organizing an information and communications technology road show, regulatory workshop series, and broadband innovation information exchanges; developing an agenda for cooperation on intelligent transportation and freight systems; making more efficient use of the North

American Development Bank and the Border Environment Cooperation Commission; and pursue joint investment initiatives.[14]

High-Level Regulatory Cooperation Council

Another bilateral effort is the U.S.-Mexico High-Level Regulatory Cooperation Council (HLRCC) launched in May 2010. The official work plan was released by the two governments on February 28, 2012, and focuses on regulatory cooperation in numerous sectoral issues including: food safety; e-certification for plants and plant products; commercial motor vehicle safety standards and procedures; nanotechnology; e-health; and offshore oil and gas development standards. U.S. agencies that are involved in regulatory cooperation include the U.S. Food and Drug Administration, Department of Agriculture, Department of Transportation, Office of Management and Budget, Department of Interior, and Occupational Safety and Health Administration.[15]

21st Century Border Management

The United States and Mexico are engaged in a bilateral border management initiative under the Declaration Concerning the 21st Century Border Management that was announced in 2010. This initiative is a bilateral effort to manage the 2,000-mile U.S.-Mexico border through the following cooperative efforts: expediting legitimate trade and travel; enhancing public safety; managing security risks; engaging border communities; and setting policies to address possible statutory, regulatory, and/or infrastructure changes that would enable the two countries to improve collaboration.[16] With respect to port infrastructure, the initiative specifies expediting legitimate commerce and travel through investments in personnel, technology, and infrastructure.[17] The two countries established a Bilateral Executive Steering Committee (ESC) composed of representatives from the appropriate federal government departments and offices from both the United States and Mexico. For the United States, this includes representatives from the Departments of State, Homeland Security, Justice, Transportation, Agriculture, Commerce, Interior, Defense, and the Office of the United States Trade Representative. For Mexico, it includes representatives from the Secretariats of Foreign Relations, Interior, Finance and Public Credit, Economy, Public Security, Communications and Transportation, Agriculture, and the Office of the Attorney General of the Republic.[18]

North American Leaders Summits

The United States, Mexico, and Canada have made efforts since 2005 to increase cooperation on security and economic issues through various endeavors, most notably by participating in trilateral summits known as the North American Leaders Summits. The most recent summit took place on February 19, 2014 in Toluca, Mexico, with an agenda focused on immigration, energy, and commerce. The 2015 Summit has been postponed and is scheduled to take place later this year in Canada. Some observers view the initiatives as constructive to addressing issues of mutual interest and benefit for all three countries. Critics of the most recent summit contend that the agenda did not include human rights issues or discussions on the drug-related violence in Mexico.

During the February 2014 Summit, President Obama, Mexican President Peña Nieto, and Canadian Prime Minister Stephen Harper announced initiatives regarding the economic prosperity of the region; education initiatives; energy and climate change; citizen security; and regional, global, and stakeholder outreach.[19] The leaders discussed numerous economic and security initiatives for North America in the 21st century with the goal of setting new global standards for trade, education, sustainable growth, and innovation. In the areas of economic cooperation, discussions included developing a North American Transportation Plan; streamlining procedures and harmonizing customs data requirements; facilitating the movement of people through the establishment in 2014 of a North American Trusted Traveler Program, which will recognize and build upon existing programs; promoting trilateral exchanges on logistics corridors and regional development; and continuing prior initiatives such as protecting and enforcing intellectual property rights. In energy cooperation, the leaders continued their commitment to developing and securing affordable, clean and reliable energy supplies to help drive economic growth and support sustainable development. The leaders committed to continuing cooperation on climate change and environmental cooperation; security; and effective information exchanges and coordination among law-enforcement authorities to counter drug trafficking, arms trafficking, money laundering, and other illicit activities. The three governments also stated that they share a commitment to combating human trafficking in all its forms and agreed to work toward improving services for the victims of this crime.[20]

THE MEXICAN ECONOMY

Mexico's economy is closely linked to the U.S. economy due to the strong trade and investment ties between the two countries. After modest economic growth in 2014, the Mexican economy is expected to improve in 2015, partly because of recent reform measures but also because the outlook for the U.S. economy is positive. However, violence in Mexico remains an issue and security risks could threaten economic growth.

Economic Trends

After modest growth in 2014, economic conditions in Mexico are forecast to improve slightly in 2015. The Mexican economy performed poorly in 2014, expanding by just 2.1%. Although it performed better than in 2013, the growth rate was relatively modest and below expectations. Over the past 30 years (1984-2014), Mexico has had a low economic growth record with an average growth rate of 2.6%.[21] During the term of former president Vicente Fox (2001-2006), the average growth rate was 2.4%, and during the term of the next president, Felipe Calderon (2007- 2012), average growth rate was even lower, at 1.9%. Economic volatility during these periods of time may have affected Mexico's ability to expand at a faster right. Events such as the U.S. recession of 2001 and the global economic downturn of 2009 adversely affected the economy and offset Mexico's efforts to improve macroeconomic management. Factors such as plentiful natural resources, a young and abundant labor force, and proximity to the United States may help Mexico's economy to grow at a faster rate over the next 20 years. However, violence and security risks may hinder economic growth.[22]

Trends in Mexico's GDP growth generally follow U.S. economic trends, as shown in *Figure 2*, but with higher fluctuations. Mexico's economy is highly dependent on manufacturing exports to the United States and approximately 80% of Mexico's exports are destined for the United States. After modest GDP growth of 2.1% in 2014, Mexico's GDP growth is forecast to grow 2.7% to 3.2% in 2015 and around 3.5% in 2016.[23] The country's outlook will likely remain closely tied to that of the United States, despite Mexico's efforts to diversify trade.

Another factor affecting the economy is the price of oil. The recent sharp decline in oil prices may have an adverse effect on Mexico's economy. Oil revenues make up almost one-third of Mexico's budget and the fall in oil prices will likely result in fiscal constraints. The Mexican government has various mitigation strategies in place for 2015, but it will probably have to make adjustments over the year, especially if oil prices remain low.[24]

The government of Mexico has put forth a bold package of structural reforms to help reverse years of slow economic growth, high levels of labor market informality, and increasing income inequality. After moving forward on most of his structural reform agenda in 2013-2014, President Enrique Peña Nieto is likely to focus on the implementation of these reforms, as well as on other efforts to boost economic growth, create more jobs, and eliminate drug-related violence.[25] The reform package comes at a time of declining oil prices that may limit the government's ability to gain immediate benefits in the economy. Mexico will likely need to focus on full implementation of the reforms in the near future and combine these efforts with other actions. A 2015 OECD report suggests that Mexico will need to enact complementary reforms to address issues such as perceived corruption, weak administrative governance, and lack of judicial enforcement in order to achieve potential economic growth.[26]

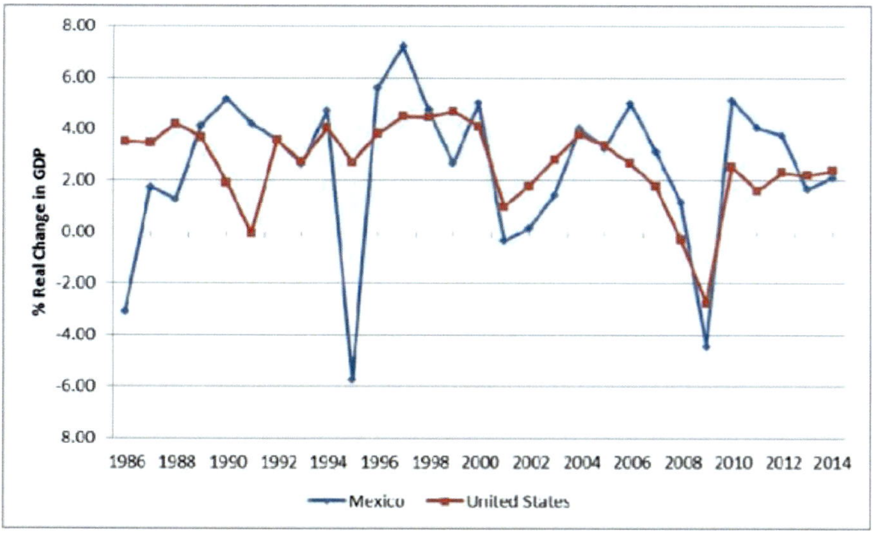

Source: Prepared by CRS using data from the Economist Intelligence Unit.

Figure 2. GDP Growth Rates for the United States and Mexico.

Informality and Poverty

Part of the government's reform efforts are aimed at making economic growth more inclusive, reducing income inequality, improving the quality of education, and reducing informality and poverty. Mexico has a large informal sector that is estimated to account for a considerable portion of total employment.

Estimates on the size of the informal labor sector vary widely, with some sources estimating that the informal sector accounts for about one-third of total employment and others estimating it to be as high as two-thirds of the workforce. Under Mexico's legal framework, workers in the formal sector are defined as salaried workers employed by a firm that registers them with the government and are covered by Mexico's social security programs. Informal sector workers are defined as non-salaried workers who are usually self-employed. These workers have various degrees of entitlement to other social protection programs. Salaried workers can be employed by industry, such as construction, agriculture, or services.

Non-salaried employees are defined by exclusion and can be defined by various categories. These workers may include agricultural producers; seamstresses and tailors; artisans; street vendors; individuals who wash cars on the street; and other professions.

Many workers in the informal sector suffer from poverty, which has been one of Mexico's more serious and pressing economic problems for many years. Although the government has made progress in poverty reduction efforts, poverty continues to be a basic challenge for the country's development. The Mexican government's efforts to alleviate poverty have focused on conditional cash transfer programs. The *Prospera* (previously called *Oportunidades*) program seeks to not only alleviate the immediate effects of poverty through cash and in-kind transfers, but to break the cycle of poverty by improving nutrition and health standards among poor families and increasing educational attainment.

According to the World Bank, *Prospera* has benefitted nearly six million families and has been replicated in 52 countries.[27]The program provides cash transfers to families in poverty who demonstrate that they regularly attend medical appointments and can certify that children are attending school. The government also provides educational cash transfers to participating families. Programs also provide nutrition support to pregnant and nursing women and malnourished children.[28]

Some economists cite the informal sector as a hindrance to the country's economic development. Other experts contend that Mexico's social programs benefitting the informal sector have led to increases in informal employment. A 2012 report by the Migration Policy Institute contends that there are two lines of argument that attempt to explain the reason for such a large informal sector: (1) overregulation of businesses; and (2) an unintended incentive to informality created by Mexico's social protection programs.[29] The report cites evidence suggesting that the scale of informality in Mexico may result in a lower level of productivity, but it is not clear whether it hinders economic growth.[30]

Structural and Other Economic Challenges

For years, numerous political analysts and economists have agreed that Mexico needs significant political and economic structural reforms to improve its potential for long-term economic growth. The Mexican government implemented numerous reform measures after the 1995 currency crisis that helped the country modify its macroeconomic policies and restore policy credibility.

Key reforms included measures to reduce public debt, the introduction of a balanced budget rule, an inflation targeting framework and a floating exchange rate policy. Such policies positioned the country well in terms of macroeconomic and financial performance, but economic growth remains insufficient and many experts agree that more needs to be done to improve well-being in all regions of the country.

Much credit has been given to President Peña Nieto for being able to break gridlock in the Mexican government and help pass reform measures that are expected to stimulate economic growth. In its 2015 economic survey for Mexico, the OECD states that Mexico is its top reformer over the past two years and deserves acclaim.[31]

The study says that the main challenge for the government now is to ensure full implementation of these reforms and that it must progress further in other key areas that have not been tackled. For the reforms to be fully implemented, according to the study, Mexico must improve administrative capacity at all levels of government and reform its judicial institutions. The study contends that such actions have a strong potential to boost living standards substantially, stimulate economic growth, and reduce income inequality.[32]

Issues regarding human rights conditions, rule of law, and corruption are also challenges that need to be addressed by the government as they too affect economic conditions and living standards.

U.S. policymakers have expressed ongoing concerns about these issues and may take an interest in how well President Peña Nieto is implementing judicial reforms.[33]

According to a 2014 study by the McKinsey Global Institute, Mexico has been successful in creating certain sectors of the economy that are highly competitive in the world market, but has not been so successful in others.[34] The study describes a "dualistic" nature of the Mexican economy in which there is a modern Mexico with sophisticated automotive and aerospace factories, multinationals that compete in global markets, and universities that graduate high numbers of engineers.

In contrast, the other part of Mexico, according to the study, is technologically backward, unproductive, and operates outside the formal economy.[35] The study states that three decades of economic reforms have failed to raise the overall GDP growth. Government measures to privatize industries, liberalize trade, and welcome foreign investment have created a side to the economy that is highly productive in which numerous industries have flourished, but the reforms have not been successful in touching other sectors of the economy where traditional enterprises have not modernized, informality is rising, and productivity is plunging.[36]

Energy Sector

Mexico's long-term economic outlook depends largely on the energy sector. [37] Mexico is one of the 10 largest oil producers in the world, and is the third-largest in the Western Hemisphere. However, Mexico's oil production has steadily decreased since 2005 as a result of natural production declines.[38] The oil sector generated 13% of Mexico's export earnings in 2013.[39] The Mexican government depends heavily on oil revenues, which provide 30% to 40% of the government's fiscal revenues. Many industry experts contend that Mexican oil production has peaked, and that the country's production will continue to decline in the coming years unless the Mexican government reforms its energy sector.

The Mexican government has used oil revenues from its state oil company, Pemex, for government operating expenses, which has come at the expense of needed reinvestment in the company itself.

In the final quarter of 2013, Pemex reportedly paid 50% of its revenue ($16 billion) in taxes to the federal government yet posted a total loss of $13 billion for 2013.[40] Because the government relies so heavily on oil income, any decline in revenue has major fiscal implications.

According to industry exports, Mexico has the potential resources to support a long-term recovery in total production, primarily in the Gulf of Mexico.

However, the country does not have the technical capability or financial means to develop potential deepwater projects or shale oil deposits in the north. Mexico's oil and natural gas production is unlikely to increase without improvement in Pemex's financial situation, technical abilities, and terms for investors.[41]

Energy reform is the centerpiece of the President Peña Nieto administration's attempts to overhaul the economy, attract greater foreign investment and generate more jobs. In December 2013, the Mexican President signed into law constitutional reforms related to Mexico's energy sector that aim to bolster the country's declining oil production and to allow private and foreign investment to help Pemex tap into the country's shale and deep water reserves. On August 6, 2014, Mexican lawmakers gave final approval to rules for awarding private oil contracts for the first time since 1938.

While it is difficult to predict how increasing private participation in Mexico's oil and gas sectors may affect the country's economic development, skeptics see reason to doubt the government's positive predictions. Some argue that multinational companies and large Mexican conglomerates stand more to gain from the energy reform than the Mexican people.[42] Other critics question the government's claim that the reforms will create thousands of jobs and maintain that because Pemex is a bloated company with too many employees, it would likely shed workers as a result of the reform. Others are concerned that the oil revenue will be mishandled by corrupt Pemex or government officials rather than invested in strategic ways that will benefit the country as a whole.[43]

Mexico's leftist PRD party opposed the energy reforms, arguing that reforms did not include measures to address corruption, transparency, government accountability, worker rights, and protection of affected communities and land owner rights.[44] It called for a *consulta popular*, or referendum, on the acceptance on whether the acceptance or rejection of the energy reform is constitutionally permissible. Mexico's Supreme Court, however, blocked this referendum.

In February 2012, the United States and Mexico signed the U.S.-Mexico Trans-Boundary Hydrocarbon agreement, which addresses issues related to the development of oil and gas reservoirs that cross the international maritime boundary between the two countries in the Gulf of Mexico. After review by the U.S. Congress, the agreement was approved as part of the Bipartisan Budget Act of 2013 (P.L. 113-67). According to the U.S. Department of the Interior, the agreement involves two U.S. actions: lifting a moratorium and jointly developing resources in a "transboundary area"—areas straddling the U.S.-Mexico marine border. The Obama Administration stated that the agreement was a step forward in clarifying relations between the two countries in managing energy resources in portions of the Gulf of Mexico and also represented an example of U.S.-Mexico efforts to develop a sustainable energy trade relationship.[45]

Mexico's Regional Free Trade Agreements

Mexico has had a growing commitment to trade integration and liberalization through the formation of free trade agreements (FTAs) since the 1990s, and its trade policy is among the most open in the world. The pursuit of FTAs with other countries not only provides economic benefits, but could also potentially reduce Mexico's economic dependence on the United States. In an effort to increase trade with other countries, Mexico has a total of 12 free trade agreements involving 44 countries. These include agreements with most countries in the Western Hemisphere, including the United States and Canada under NAFTA, Chile, Colombia, Costa Rica, Nicaragua, Peru, Guatemala, El Salvador, and Honduras.

Mexico has ventured out of the hemisphere in negotiating FTAs, and, in July 2000, entered into agreements with Israel and the European Union. Mexico became the first Latin American country to have preferred access to these two markets. Mexico has also completed an FTA with the European Free Trade Association (EFTA) of Iceland, Liechtenstein, Norway, and Switzerland.

The Mexican government has continued to look for potential free trade partners, and expanded its outreach to Asia in 2000 by entering into negotiations with Singapore, Korea, and Japan. Negotiations on FTAs with Korea and Singapore are stalled. In addition to the bilateral and multilateral free trade agreements, Mexico is a member of the WTO,[46] the Asia-Pacific Economic Cooperation (APEC) forum, and the OECD.

Mexico is a member country of the Pacific Alliance, a regional trade integration initiative formed by Chile, Colombia, Mexico, and Peru on April 11, 2011. Its main purpose is for members to form a regional trading bloc and forge stronger economic ties with the Asia-Pacific region. The Alliance is not an FTA, but it is intended to supplement existing FTAs among member countries. The concept is for member countries to act as a unified economic bloc to negotiate and trade with other countries.[47]

NAFTA

The North American Free Trade Agreement (NAFTA) has been in effect since January 1994.[48] The overall economic impact of NAFTA is difficult to measure since trade and investment trends are influenced by numerous other economic variables such as economic growth, inflation, and currency fluctuations.

The agreement may have accelerated the trade liberalization that was already taking place between the United States and Mexico, but many of these changes may have taken place with or without an agreement. Nevertheless, NAFTA is significant because it was the most comprehensive free trade agreement (FTA) negotiated at the time, and contained several groundbreaking provisions. There are numerous indications that NAFTA has achieved many of the intended trade and economic benefits, as well as incurred adjustment costs. This has been in keeping with what most economists maintain, that trade liberalization promotes overall economic growth among trading partners, but that there are significant adjustment costs.

Most of the trade effects in the United States related to NAFTA are due to changes in U.S. trade and investment patterns with Mexico. At the time of NAFTA implementation, the U.S.-Canada Free Trade Agreement already had been in effect for five years, and some industries in the United States and Canada were already highly integrated. Mexico, on the other hand, had followed an aggressive import-substitution policy for many years prior to NAFTA in which it had sought to develop certain domestic industries through trade protection. One example is the Mexican automotive industry, which had been regulated by a series of five decrees issued by the Mexican government between 1962 and 1989. The decrees established import tariffs as high as 25% on automotive goods and had high restrictions on foreign auto production in Mexico. Under NAFTA, Mexico agreed to eliminate these restrictive trade policies.

Prior to NAFTA, Mexico was already liberalizing its protectionist trade and investment policies that had been in place for decades. The restrictive trade regime began after Mexico's revolutionary period, and remained until the early to mid-1980s, when it began to shift to a more open, export-oriented economy. For Mexico, an FTA with the United States represented a way to lock in the trade reforms, attract greater flows of foreign investment, and spur economic growth. For the United States, NAFTA represented an opportunity to expand the growing export market to the south, but it also represented a political opportunity to improve the relationship with Mexico.

Estimating the economic impact of trade agreements is very difficult due to a lack of data and important theoretical and practical matters associated with generating results from economic models. In addition, such estimates provide an incomplete accounting of the total economic effects of trade agreements.[49] Numerous studies suggest that NAFTA achieved many of the intended trade and economic benefits.[50] Other studies suggest that NAFTA has come at a cost to U.S. workers.[51] This has been in keeping with what most economists maintain, that trade liberalization promotes overall economic growth among trading partners, but that there are both winners and losers from adjustments.

Not all changes in trade and investment patterns within North America since 1994 can be attributed to NAFTA because trade has also been affected by a number of factors. The sharp devaluation of the peso at the end of the 1990s and the associated recession in Mexico had considerable effects on trade, as did the rapid growth of the U.S. economy during most of the 1990s and, more recently, the economic slowdown caused by the 2008 financial crisis. Trade-related job gains and losses since NAFTA may have accelerated trends that were ongoing prior to NAFTA and may not be totally attributable to the trade agreement.

Many economists and other observers have credited NAFTA with helping U.S. manufacturing industries, especially the U.S. auto industry, become more globally competitive through the development of supply chains.[52] Much of the increase in U.S.-Mexico trade, for example, can be attributed to specialization as manufacturing and assembly plants have reoriented to take advantage of economies of scale. As a result, supply chains have been increasingly crossing national boundaries as manufacturing work is performed wherever it is most efficient.[53] A reduction in tariffs in a given sector not only affects prices in that sector but also in industries that purchase intermediate inputs from that sector. The expansion of trade resulted in the creation of vertical supply relationships, especially along the U.S.-Mexico border.

The flow of intermediate inputs produced in the United States and exported to Mexico and the return flow of finished products greatly increased the importance of the U.S.-Mexico border region as a production site. U.S. manufacturing industries, including automotive, electronics, appliances, and machinery, all rely on the assistance of Mexican manufacturers. One study estimates that 40% of the content of U.S. imports from Mexico and 25% of the content of U.S. imports from Canada are of U.S. origin. In comparison, U.S. imports from China are said to have only 4% U.S. content. Taken together, goods from Mexico and Canada represent about 75% of all the U.S. domestic content that returns to the United States as imports.[54]

PROPOSED TRANS-PACIFIC PARTNERSHIP (TPP) AGREEMENT

The Trans-Pacific Partnership (TPP) is a proposed regional free trade agreement (FTA) being negotiated among the United States, Australia, Brunei, Canada, Chile, Japan, Malaysia, Mexico, New Zealand, Peru, Singapore, and Vietnam. U.S. negotiators and others describe and envision the TPP as a "comprehensive and high-standard" FTA that aims to liberalize trade in nearly all goods and services and include rules-based commitments beyond those currently established in the World Trade Organization (WTO). Mexico and Canada began participating in the TPP negotiations in December 2012. If concluded as envisioned, the TPP potentially could eliminate tariff and nontariff barriers to trade and investment among the parties and could serve as a template for a future trade pact among APEC members and potentially other countries. Congress has a direct interest in the negotiations, both through influencing U.S. negotiating positions with the executive branch, and by considering legislation to implement any resulting agreement.[55]

The TPP negotiations are reportedly nearing conclusion. As the negotiations are confidential, it is difficult to assess how many issues remain unresolved.

Remaining issues are likely also the most sensitive and their resolution may require high-level political decisions. Conclusion of the TPP negotiations may also be impacted by congressional consideration of Trade Promotion Authority (TPA), legislation which some Members of Congress view as necessary before TPP is concluded and then considered by Congress.[56]

The proposed TPP would likely enhance the links Mexico already has with the United States and Canada under NAFTA. The Mexican government agreed to several conditions that TPP countries had placed on its entry into the negotiations, including a commitment to "high standards." The conditions included that Mexico would not be able to reopen any existing agreements that were already made by the current TPP partners, unless they agreed to revisit something previously agreed upon. The government of Mexico is seeking to have a unified vision with NAFTA partners by negotiating as a North American region in the TPP talks. Mexico is interested in ensuring the TPP addresses energy trade issues; drops foreign trade barriers in the auto sector, especially in Japan; expands access for the Mexican aerospace industry; and protects the textile industry and shoe makers from further competition from Asia. In agriculture, Mexico is unlikely to raise specific concerns as it has already undertaken significant reform and market opening measures through NAFTA and other unilateral actions. NAFTA will reportedly continue to co-exist with a TPP so that Mexican exporters could continue to export through NAFTA.[57]

BILATERAL TRADE ISSUES

The United States and Mexico have had a number of trade disputes over the years, many of which have been resolved. These issues have involved trade in sugar, country of origin labeling, tomato imports from Mexico, dolphin-safe tuna labeling, and NAFTA trucking provisions.

Sugar Disputes

2014 Mexican Sugar Import Dispute

On December 19, 2014, the U.S. Department of Commerce (DOC) signed an agreement with the Government of Mexico suspending the U.S. countervailing duty (CVD) investigation of sugar imports from Mexico. The DOC signed a second agreement with Mexican sugar producers and exporters suspending an antidumping (AD) duty investigation on imports of Mexican sugar. The agreements suspending the investigations alter the nature of trade in sugar between Mexico and the United States by 1) imposing volume limits on U.S. sugar imports from Mexico, and 2) setting minimum price levels on Mexican sugar.[58]

After the suspension agreement was announced, two U.S. sugar companies, Imperial Sugar Company and AmCane Sugar LLC, requested that the DOC continue the CVD and AD investigations on sugar imports from Mexico. The two companies filed separate submissions on January 16, 2015 claiming "interested party" status. The companies claimed they met the statutory standards to seek continuation of the probes. The submissions to the DOC followed requests to the ITC, by the same two companies, to review the two December 2014 suspension agreements.[59] The ITC reviewed the sugar suspension agreements to determine whether they completely eliminate the injurious effect of sugar imports from Mexico. On March 19, 2015, the ITC upheld the agreement between the United States and Mexico that suspended the sugar investigations. Mexican Economy Minister Ildefonso Guajardo Villarreal praised the ITC decision, stating that it supported the Mexican government position.[60]

The dispute began on March 28, 2014, when the American Sugar Coalition and its members filed a petition requesting that the U.S. ITC and the DOC conduct an investigation alleging that Mexico was dumping and subsidizing its sugar exports to the United States. The petitioners claimed that dumped and subsidized sugar exports from Mexico were harming U.S. sugar producers and workers. They claimed that Mexico's actions would cost the industry $1 billion in 2014. On April 18, 2014, the DOC announced the initiation of AD and CVD investigations of sugar imports from Mexico.[61] On May 9, 2014, the ITC issued a preliminary report stating that there was a reasonable indication a U.S. industry was materially injured by imports of sugar from Mexico that were allegedly sold in the United States at less than fair value and allegedly subsidized by the Government of Mexico.[62]

In August 2014, the DOC announced in its preliminary ruling that Mexican sugar exported to the United States was being unfairly subsidized. Following the preliminary subsidy determination, the DOC stated that it would direct the U.S. Customs and Border Protection to collect cash deposits on imports of Mexican sugar. Based on the preliminary findings, the DOC imposed cumulative duties on U.S. imports of Mexican sugar, ranging from 2.99% to 17.01% under the CVD order. Additional duties of between 39.54% and 47.26% were imposed provisionally following the preliminary AD findings.[63] The final determination in the two investigations was expected in later in 2015 and had not been issued when the suspension agreements were signed.

The Sweetener Users Association (SUA), which represents beverage makers, confectioners, and other food companies, argues that the case is "a diversionary tactic to distract from the real cause of distortion in the U.S. sugar market—the U.S. government's sugar program."[64] It contends that between 2009 and 2012, U.S. sugar prices soared well above the world price because of the U.S. program, providing an incentive for sugar growers to increase production. According to the sugar users association, this resulted in a surplus of sugar and a return to lower sugar prices.[65] The SUA has been a critic of the U.S. sugar program.[66]

Sugar and High Fructose Corn Syrup Dispute Resolved in 2006

In 2006, the United States and Mexico resolved a previous trade dispute involving sugar and high fructose corn syrup. The dispute involved a sugar side letter negotiated under NAFTA. Mexico argued that the side letter entitled it to ship net sugar surplus to the United States duty-free under NAFTA, while the United States argued that the sugar side letter limited Mexican shipments of sugar. In addition, Mexico complained that imports of high fructose corn syrup (HFCS) sweeteners from the United States constituted dumping. It imposed anti-dumping duties for some time, until NAFTA and WTO dispute resolution panels upheld U.S. claims that the Mexican government colluded with the Mexican sugar and sweetener industries to restrict HFCS imports from the United States.

In late 2001, the Mexican Congress imposed a 20% tax on soft drinks made with corn syrup sweeteners to aid the ailing domestic cane sugar industry, and subsequently extended the tax annually despite U.S. objections. In 2004, the United States Trade Representative (USTR) initiated WTO dispute settlement proceedings against Mexico's HFCS tax, and following interim decisions, the WTO panel issued a final decision on October 7, 2005, essentially supporting the U.S. position. Mexico appealed this decision, and in March 2006, the WTO Appellate Body upheld its October 2005 ruling. In July 2006, the United States and Mexico agreed that Mexico would eliminate its tax on soft drinks made with corn sweeteners no later than January 31, 2007. The tax was repealed, effective January 1, 2007.

The United States and Mexico reached a sweetener agreement in August 2006. Under the agreement, Mexico can export 500,000 metric tons of sugar duty-free to the United States from October 1, 2006, to December 31, 2007. The United States can export the same amount of HFCS duty-free to Mexico during that time. NAFTA provides for the free trade of sweeteners beginning January 1, 2008. The House and Senate sugar caucuses expressed objections to the agreement, questioning the Bush Administration's determination that Mexico is a net-surplus sugar producer to allow Mexican sugar duty-free access to the U.S. market.[67]

Country-of-Origin Labeling (COOL)

The United States has been involved in a country-of-origin labeling (COOL) trade dispute with Canada and Mexico for several years. Mexican and Canadian meat producers claim that U.S. mandatory COOL requirements for animal products discriminate against their products. They contend that the labeling requirements create an incentive for U.S. meat processors to use exclusively domestic animals because it forces processors to segregate animals born in Mexico or Canada from U.S.-born animals, which is very costly. They state that the COOL requirement is an unfair barrier to trade. A World Trade Organization (WTO) appellate panel in June 2013 ruled against the United States. The United States appealed the decision. The WTO appellate body decision on U.S. compliance is due by May 18, 2015. If the appellate body rules against the United States, Mexico has stated that it would consider imposing retaliatory tariffs on a wide variety of U.S. exports to Mexico, including fruits and vegetables, juices, meat products, dairy products, machinery, furniture and appliances, and others.[68] Some U.S. policymakers and meat producers support repealing U.S. COOL requirements on meat products to avoid retaliatory trade measures.[69]

The dispute began on December 1, 2008, when Canada requested WTO consultations with the United States concerning certain mandatory labeling provisions required by the 2002 farm bill (P.L. 107-171) as amended by the 2008 farm bill (P.L. 110-246). On December 12, 2008, Mexico requested to join the consultations. U.S. labeling provisions include the obligation to inform consumers at the retail level of the country of origin in certain commodities, including beef and pork.[70]

The U.S. Department of Agriculture (USDA) labeling rules for meat and meat products have been controversial. A number of livestock and food industry groups oppose COOL as costly and unnecessary. Canada and Mexico, the main livestock exporters to the United States, argue that COOL has a trade-distorting impact by reducing the value and number of cattle and hogs shipped to the U.S. market, thus violating WTO trade commitments. Others, including some cattle and consumer groups, maintain that Americans want and deserve to know the origin of their foods.[71]

In November 2011, the WTO dispute settlement panel found that 1) COOL treats imported livestock less favorably than U.S. livestock and 2) that it does not meet its objective to provide complete information to consumers on the origin of meat products. In March 2012, the United States appealed the WTO ruling. In June 2012, the WTO's Appellate Body upheld the finding that

COOL treats imported livestock less favorably than domestic livestock and reversed the finding that it does not meet its objective to provide complete information to consumers. It could not determine if COOL was more trade restrictive than necessary.

In order to meet a compliance deadline by the WTO, USDA issued a revised COOL rule on May 23, 2013, that required meat producers to specify on retail packaging where each animal was born, raised, and slaughtered; and that prohibited the mixing of muscle cuts from different countries. Canada and Mexico challenged the 2013 labeling rules before a WTO compliance panel. The compliance panel sided with Canada and Mexico; the United States is appealing the decision.[72] If the Appellate Body upholds the compliance panel findings, and the United States does not alter the COOL law to bring it into compliance with WTO obligations, Mexico is expected to request WTO authorization to retaliate against the United States with increased tariffs on imports of various U.S. products.

Mexican Trucking Issue

On January 9, 2015, the Department of Transportation's Federal Motor Carrier Safety Administration (FMCSA) announced that Mexican motor carriers would soon be able to apply for authority to conduct long-haul, cross-border trucking services in the United States, marking a significant milestone in implementation of U.S. NAFTA commitments.[73] The International Brotherhood of Teamsters filed a lawsuit on March 20, 2015, in the U.S. Court of Appeals for the Ninth Circuit, seeking to halt FMCSA's move to allow Mexican motor carriers to operate in the United States. The Mexican government stated that it would consider retaliatory measures if the Teamsters lawsuit is successful.[74]

The implementation of NAFTA trucking provisions has been a major trade issue between the United States and Mexico since 2000. The United States had delayed its trucking commitments under NAFTA for many years. It was to have given Mexican commercial trucks full access to four U.S.-border states in 1995 and full access throughout the United States in 2000. Citing safety concerns, the United States did not implement NAFTA's trucking provisions.[75] The Mexican government objected and claimed that U.S. actions were a violation of U.S. commitments under NAFTA. A dispute resolution panel supported Mexico's position in February 2001. President Bush indicated a willingness to implement the provision, but the U.S. Congress required additional safety provisions in the FY2002 Department of Transportation Appropriations Act (P.L. 107-87). The United States and Mexico cooperated to resolve the issue over the years and engaged in numerous talks regarding safety and operational issues. The United States had two pilot programs on cross-border trucking to help resolve the issue: the Bush Administration's pilot program of 2007 and the Obama Administration's program of 2011.

The Obama Administration's 2011 program expired in October 2014. After the January 2015 FMCSA announcement, Department of Transportation Secretary Anthony Foxx stated that "opening the door to a safe cross-border trucking system with Mexico" would strengthen relations and meet U.S. obligations under NAFTA.[76] The FMCSA submitted a report to Congress in which it states that the Obama Administration pilot program successfully demonstrated that Mexican motor carriers operate throughout the United States at a safety level equivalent to U.S. and Canada-domiciled motor carriers.[77] The DOT is expected to begin taking applications after the publication of a Federal Register notice. The International Brotherhood of Teamsters has repeatedly characterized the pilot program as a failure.[78]

Bush Administration's Pilot Program of 2007

On November 27, 2002, with safety inspectors and procedures in place, the Bush Administration announced that it would begin the process that would open U.S. highways to Mexican truckers and buses. However, environmental and labor groups went to court in early December to block the action. On January 16, 2003, the U.S. Court of Appeals for the Ninth Circuit ruled that full environmental impact statements were required for Mexican trucks to be allowed to operate on U.S. highways. The U.S. Supreme Court reversed that decision on June 7, 2004.

In February 2007, the Bush Administration announced a pilot project to grant Mexican trucks from 100 transportation companies full access to U.S. highways. In September 2007, the Department of Transportation (DOT) launched a one-year pilot program to allow approved Mexican carriers beyond the 25-mile commercial zone in the border region, with a similar program allowing U.S. trucks to travel beyond Mexico's border and commercial zone. Over the 18 months that the program existed, 29 motor carriers from Mexico were granted operating authority in the United States. Two of these carriers dropped out of the program shortly after being accepted, while two others never sent trucks across the border. In total, 103 Mexican trucks were used by the carriers as part of the program.[79]

In the FY2008 Consolidated Appropriations Act (P.L. 110-161), signed into law in December 2007, Congress included a provision prohibiting the use of FY2008 funding for the establishment of the pilot program. However, the DOT determined that it could continue with the pilot program because it had already been established. In March 2008, the DOT issued an interim report on the cross-border trucking demonstration project to the Senate Committee on Commerce, Science, and Transportation. The report made three key observations: (1) the Federal Motor Carrier Safety Administration (FMCSA) planned to check every participating truck each time it crossed the border to ensure that it met safety standards; (2) there was less participation in the project than was expected; and (3) the FMCSA implemented methods to assess possible adverse safety impacts of the project and to enforce and monitor safety guidelines.[80]

In early August 2008, DOT announced that it would be extending the pilot program for an additional two years. In opposition to this action, the House approved on September 9, 2008 (by a vote of 396 to 128), H.R. 6630, a bill that would have prohibited DOT from granting Mexican trucks access to U.S. highways beyond the border and commercial zone. The bill also would have prohibited DOT from renewing such a program unless expressly authorized by Congress. No action was taken by the Senate on the measure.

On March 11, 2009, the FY2009 Omnibus Appropriations Act (P.L. 111-8) terminated the pilot program. The FY2010 Consolidated Appropriations Act, passed in December 2009 (P.L. 111- 117), did not preclude funds from being spent on a long-haul Mexican truck pilot program, provided that certain terms and conditions were satisfied. Numerous Members of Congress urged President Obama to find a resolution to the dispute in light of the effects that Mexico's retaliatory tariffs were having on U.S. producers (see section below on President Obama's program).

Mexico's Retaliatory Tariffs of 2009 and 2010

In response to the abrupt end of the pilot program, the Mexican government retaliated in 2009 by increasing duties on 90 U.S. products with a value of $2.4 billion in exports to Mexico. Mexico began imposing tariffs in March 2009 and, after reaching an understanding with the United States, eliminated them in two stages in 2011. The retaliatory tariffs ranged from 10% to 45% and covered a range of products that included fruit, vegetables, home appliances, consumer products, and paper.[81] Subsequently, a group of 56 Members of the House of Representatives wrote to United States Trade Representative Ron Kirk and DOT Secretary Ray LaHood requesting the

Administration to resolve the trucking issue.[82] The bipartisan group of Members stated that they wanted the issue to be resolved soon because the higher Mexican tariffs were having a "devastating" impact on local industries, especially in agriculture, and area economies in some states. One reported estimate stated that U.S. potato exports to Mexico had fallen 50% by value since the tariffs were imposed and that U.S. exporters were losing market share to Canada.[83]

On August 16, 2010, the Mexican government announced a revised list of retaliatory tariffs on imports from the United States. The revised list added 26 products to and removed 16 products from the original list of 89, bringing the new total to 99 products from 43 states with a total export value of $2.6 billion. Products that were added to the list included several types of pork products, several types of cheeses, sweet corn, pistachios, oranges, grapefruits, apples, oats and grains, chewing gum, ketchup, and other products. The largest in terms of value were two categories of pork products, which had an estimated export value of $438 million in 2009. Products that were removed from the list included peanuts, dental floss, locks, and other products.[84] The revised retaliatory tariffs were lower than the original tariffs and ranged from 5% to 25%. Mexico reportedly rotated the list of products to put more pressure on the United States to seek a settlement for the trucking dispute.[85] U.S. producers of fruits, pork, cheese, and other products that were bearing the cost of the retaliatory tariffs reacted strongly at the lack of progress in resolving the trucking issue and argued, both to the Obama Administration and to numerous Members of Congress, that they were potentially losing millions of dollars in sales as a result of this dispute.

In March 2011, President Obama and Mexican President Calderón announced that they had agreed on a way to move forward to resolving the dispute. Mexico stated that once a final agreement was reached, it would suspend retaliatory tariffs in stages, beginning with reducing tariffs by 50% at the signing of an agreement and suspending the remaining 50% when the first Mexican carrier was granted operating authority under the program.[86] By October 2011, Mexico had suspended all retaliatory tariffs on U.S. exports to Mexico.

The Obama Administration's Trucking Program of 2011

In January 2011, the Obama Administration presented an "initial concept document" to Congress and the Mexican government for a new long-haul trucking pilot program with numerous safety inspection requirements for Mexican carriers. The concept document would put in place a new inspection and monitoring regime in which Mexican carriers would have to apply for long-haul operating authority. The project would include several thousand trucks and eventually bring as many vehicles as are needed into the United States.[87] A DOT press release from January 6, 2011, stated that a formal proposal on which the public would have the opportunity to comment would be released in the following months.[88] The Mexican government responded positively to the initiative, stating that it would not continue rotating the list of retaliatory tariffs, but that it would keep the current tariffs in place until a final accord was reached.[89]

The concept document outlined three sets of elements. The first set of elements, pre-operations elements, included an application process for Mexican carriers interested in applying for long-haul operations in the United States; a vetting process by the U.S. Department of Homeland Security and the Department of Justice; a safety audit of Mexican carriers applying for the program; documentation of Mexican commercial driver's license process to demonstrate comparability to the U.S. process; and evidence of financial responsibility (insurance) of the applicant. The second set of elements, operations elements, included the following: monitoring procedures that included regular inspections and electronic monitoring of long-haul vehicles and drivers; a follow-up review (first review) to ensure continued safe operation; a compliance review (second review) upon which a participating carrier would be eligible for full operation authority; and a FMCSA review that included insurance monitoring and drug and alcohol collection and testing facilities. The third set of elements, transparency elements, would require Federal Register notices by the FMCSA; a publically accessible website that provides information on participating carriers; the establishment of a Federal Advisory Committee with representation from a diverse group of stakeholders; periodic reports to Congress; and requirements for DOT Office of the Inspector General reports to Congress.[90]

2011 Memorandum of Understanding to Resolve the Dispute

On July 6, 2011, the two countries signed a Memorandum of Understanding (MOU) to resolve the dispute over long-haul cross-border trucking.[91] Within 10 days after signing of the MOU, Mexico suspended 50% of the retaliatory tariffs. Mexico agreed to suspend the remainder of the tariffs within five days of the first Mexican trucking company receiving its U.S. operating authority.[92] On October 21, 2011, Mexico suspended the remaining retaliatory tariffs.

The pilot program, which ended in October 2014, was announced by the FMCSA in July 2011. DOT stressed that roadway safety would be a priority.[93] It came as a result of numerous meetings between the Secretary of DOT, other Obama Administration officials, lawmakers, safety advocates, industry representatives, and others to address concerns. According to the FMCSA, the final text of the program addressed recommendations of over 2,000 commenters to the proposal issued in April 2011.[94] The program required trucks to comply with all Federal Motor Vehicle Safety Standards and to have electronic monitoring systems to track hours-of-service compliance. In addition, DOT reviewed the complete driving record of each driver in addition to having drug testing requirements for all drivers. Other requirements included an assessment of abilities to understand the English language and U.S. traffic signs.[95] Mexico provided reciprocal authority for U.S. carriers to engage in cross-border long-haul operations in Mexico.

On October 14, 2011, the FMCSA granted the first permit to provide international long-haul cargo services to Monterrey-based trucking firm *Transportes Olympic*. The company successfully completed a pre-authorization safety audit and had been a participant in the Bush Administration's 2007 pilot program.[96] By 2014, fifteen trucking companies from Mexico participated in the pilot program, crossing the border more than 28,000 times, traveled more than 1.5 million miles in the United States, and were subjected to 5,500 safety inspections by U.S. officials. The FMCSA stated that data collected on the pilot carriers from Mexico met the level of safety shown by American and Canadian-domiciled motor carriers.[97]

Mexican Tomatoes

In February 2013, the United States and Mexico reached an agreement on cross-border trade in tomatoes, averting a potential trade war between the two countries.[98] On March 4, 2013, the Department of Commerce (DOC) and the government of Mexico officially signed the agreement suspending the antidumping investigation on fresh tomatoes from Mexico.[99] The dispute began on June 22, 2012, when a group of Florida tomato growers, who were backed by growers in other states, asked the DOC and the U.S. International Trade Commission to terminate an antidumping duty suspension pact on tomatoes from Mexico.

The termination of the pact, which sets a minimum reference price for Mexican tomatoes in the United States, would have effectively led to an antidumping investigation on Mexican tomatoes. Mexico's Ambassador to the United States at the time, Arturo Sarukhan, warned that such an action would damage the U.S.-Mexico trade agenda and bilateral trade relationship as a whole. He also stated that Mexico would use all resources at its disposal, including the possibility of retaliatory tariffs, to defend the interests of the Mexican tomato industry.[100] The Florida Tomato Exchange, a coalition of Florida tomato growers, is challenging the suspension agreement and has a pending lawsuit filed with the U.S. Court of International Trade.[101]

The suspension pact dates back to 1996, when the DOC, under pressure from Florida tomato growers, filed an anti-dumping petition against Mexican tomato growers and began an investigation into whether they were dumping Mexican tomatoes on the U.S. market at below-market prices. NAFTA, which entered into force in January 1994, had eliminated U.S. tariffs on Mexican tomatoes, causing an inflow of fresh tomatoes from Mexico. Florida tomato growers complained that Mexican tomato growers were selling tomatoes at below-market prices.

After the 1996 filing of the petition, the DOC and Mexican producers and exporters of tomatoes reached an agreement under which Mexican tomato growers agreed to revise their prices by setting a minimum reference price in order to eliminate the injurious effects of fresh tomato exports to the United States.[102] The so-called "suspension agreement" remained in place for years and was renewed in 2002 and 2008.[103]

The 2013 suspension agreement covers all fresh and chilled tomatoes, excluding those intended for use in processing. It increases the number of tomato categories with established reference prices from one to four. It also raises reference prices at which tomatoes can be sold in the U.S. market to better reflect the changes in the marketplace since the last agreement had been signed. It continues to account for winter and summer seasons.[104]

When they filed the 2012 petition asking for the termination of the suspension agreement, U.S. tomato producers argued that the pacts had not worked. The petitioners stated that it was necessary to end the agreement with Mexico in order to "restore fair competition to the market and eliminate the predatory actions of producers in Mexico."[105] However, business groups urged the DOC to proceed cautiously in the tomato dispute since termination could result in higher tomato prices in the United States and lead Mexico to implement retaliatory measures.

Some businesses urged a continuation of the agreement, arguing that it helped stabilize the market and provide U.S. consumers with consistent and predictable pricing. According to a *New York Times* article, the Mexican tomato producers enlisted roughly 370 U.S. businesses, including Wal-Mart Stores and meat and vegetable producers, to argue their cause.[106]

Dolphin-Safe Tuna Labeling Dispute

The United States and Mexico are involved in a trade dispute under the WTO regarding U.S. dolphin-safe labeling provisions and tuna imports from Mexico. U.S. labeling provisions establish conditions under which tuna products may voluntarily be labeled as "dolphin-safe." Products may not be labeled as dolphin-safe if the tuna is caught by intentionally encircling dolphins with nets. According to the Office of the United States Trade Representative (USTR), some Mexican fishing vessels use this method when fishing for tuna. Mexico asserts that U.S. tuna labeling provisions deny Mexican tuna effective access to the U.S. market.[107] In April 2015, the date of the most recent development on the issue, a WTO compliance panel report concluded that U.S. tuna labeling rules unfairly discriminate against Mexico's fishing. Both the United States and Mexico are permitted to appeal any of the compliance panel's findings.[108]

The tuna labeling dispute began over 10 years ago. In April 2000, the Clinton Administration lifted an embargo on Mexican tuna under relaxed standards for a dolphin-safe label. This was in accordance with internationally agreed procedures and U.S. legislation passed in 1997 that encouraged the unharmed release of dolphins from nets. However, a federal judge in San Francisco ruled that the standards of the law had not been met, and the Federal Appeals Court in San Francisco sustained the ruling in July 2001. Under the Bush Administration, the Commerce Department ruled on December 31, 2002, that the dolphin-safe label may be applied if qualified observers certify that no dolphins were killed or seriously injured in the netting process. Environmental groups, however, filed a suit to block the modification. On April 10, 2003, the U.S. District Court for the Northern District of California enjoined the Commerce Department from modifying the standards for the dolphin-safe label. On August 9, 2004, the federal district court ruled against the Bush Administration's modification of the dolphin-safe standards and reinstated the original standards in the 1990 Dolphin Protection Consumer Information Act. That decision was appealed to the U.S. Ninth Circuit Court of Appeals, which ruled against the Administration in April 2007, finding that the Department of Commerce did not base its determination on scientific studies of the effects of Mexican tuna fishing on dolphins.

In late October 2008, Mexico initiated WTO dispute proceedings against the United States, maintaining that U.S. requirements for Mexican tuna exporters prevents them from using the U.S. "dolphin-safe" label for its products. The United States requested that Mexico refrain from proceeding in the WTO and that the case be moved to the NAFTA dispute resolution mechanism. According to the USTR, however, Mexico "blocked that process for settling this dispute."[109] In September 2011, a WTO panel determined that the objectives of U.S. voluntary tuna labeling provisions were legitimate and that any adverse effects felt by Mexican tuna producers were the result of choices made by Mexico's own fishing fleet and canners. However, the panel also found U.S. labeling provisions to be "more restrictive than necessary to achieve the objectives of the measures."[110] The Obama Administration appealed the WTO ruling.

On May 16, 2012, the WTO's Appellate Body overturned two key findings from the September 2011 WTO dispute panel. The Appellate Body found that U.S. tuna labeling requirements violate global trade rules because they treat imported tuna from Mexico less favorably than U.S. tuna. The Appellate Body also rejected Mexico's claim that U.S. tuna labeling requirements were more trade-restrictive than necessary to meet the U.S. objective of minimizing dolphin deaths.[111] The United States had a deadline of July 13, 2013, to comply with the WTO dispute ruling. In July 2013, the United States issued a final rule amending certain dolphin-safe labelling requirements to bring it into compliance with the WTO labeling requirements. On November 14, 2013, Mexico requested the establishment of a WTO compliance panel. On April 16, 2014, the Chair of the compliance panel announced that it expected to issue its final report to the parties by December 2014.[112]

The government of Mexico had requested the United States to broaden its dolphin-safe rules to include Mexico's long-standing tuna fishing technique. It cited statistics showing that modern equipment has greatly reduced dolphin mortality from its height in the 1960s and that its ships carry independent observers who can verify dolphin safety.[113] However, some environmental groups that monitor the tuna industry disputed claims by the Mexican government, stating that even if no dolphins are killed during the chasing and netting, some are wounded and later die. In other cases, they argued, young dolphin calves may not be able to keep pace and are separated from their mothers and later die. These groups contended that if the United States changed its labeling requirements, cans of Mexican tuna could be labeled as "dolphin-safe" when it was not. However, an industry spokesperson representing three major tuna processors in the United States, including StarKist, Bumblebee, and Chicken of the Sea, contended that U.S. companies would probably not buy Mexican tuna even if it is labeled as dolphin-safe because these companies "would not be in the market for tuna that is not caught in the dolphin-safe manner."[114]

POLICY ISSUES

U.S. policy makers are likely to closely follow trade issues regarding the TPP negotiations and regulatory cooperation with Mexico. They are also likely to follow ongoing economic reforms and policies implemented by the Peña Nieto government.

TPP Negotiations

Policy makers may consider how a TPP would affect NAFTA and U.S.-Mexico trade relations. Although nearly all U.S. trade with Mexico is now conducted duty and barrier free through NAFTA, the TPP negotiations may provide a venue for addressing issues that are not covered by NAFTA. The TPP may have implications for NAFTA in several areas, including IPR, investment, services trade, government procurement, as well as labor and environmental provisions. The provisions in more recent agreements that the United States has negotiated, such as the FTAs with Colombia and Peru, include commitments that go beyond NAFTA. If an agreement is reached on a TPP, Mexico may have to adhere to stronger and more enforceable labor and environmental provisions, stronger IPR provisions, as well as new rules on state-owned enterprises.[115]

Potential questions that Congress might consider include the following: If a TPP agreement is concluded, how would it affect U.S. economic relations with Mexico? Would NAFTA remain relevant? How would it affect bilateral trade? Would a TPP address concerns of policy makers related to the environment and worker rights? Would there be a difference in the enforcement mechanism? How would stronger IPR provisions affect U.S.-Mexico trade? How would a TPP affect jobs in the United States and Mexico?

Bilateral Economic Cooperation

Policy makers may consider issues on how the United States can improve cooperation with Mexico in the areas of border trade, transportation, competitiveness, economic growth, and security enhancement through the HLED, HLRCC, and the 21st Century Border Management programs mentioned earlier in this report. Some policy experts emphasize the importance of U.S.- Mexico trade in intermediate goods and supply chains and argue that the two governments can improve cooperation in cross-border trade and can invest more in improving border infrastructure. The increased security measures along the U.S.-Mexico-border, they argue, have resulted in a costly disruption in production chains due to extended and unpredictable wait times along the border.

Potential questions that Congress might consider include the following: How effectively has the United States pursued border initiatives with Mexico? What are the challenges facing U.S.- Mexico trade flows along the border? What steps can be taken by the two countries to improve competitiveness of industries located along the U.S.-Mexico border and elsewhere within the two countries? How successful have the United States and Mexico been in improving the flow of goods and services, while improving safety and security along the border? What have been the actual results of the initiatives that have been launched? To what extent has the emphasis on border security caused delays in border crossings or transportation of merchandise? How have recent efforts to facilitate trade affected the trade relationship?

Mexico's Economic Reforms

As Mexico moves forward with reform measures to modernize the energy sector and other parts of the economy, the overarching questions are how the reform agenda will be implemented; whether it will be implemented fully; how will it affect the U.S.-Mexico trade relationship; and whether it will be enough to drive economic growth among all sectors of the economy, increase employment in the formal sector, and bring more people out of poverty.

Potential oversight questions that Congress might consider include the following: How effectively are the Peña Nieto government and the Mexican Congress implementing economic reforms? What efforts are the United States and Mexico making on energy cooperation? To what extent will the energy reforms provide opportunities for U.S. oil companies? Will the reforms improve Mexican economic performance? What complementary measures could the Mexican government take to stimulate economic growth?

APPENDIX. MAP OF MEXICO

Figure A-1. Map of Mexico.

End Notes

[1] See CRS Report R42917, Mexico: Background and U.S. Relations, by Clare Ribando Seelke, and CRS Report R43313, Mexico's Oil and Gas Sector: Background, Reform Efforts, and Implications for the United States, coordinated by Clare Ribando Seelke.

[2] See CRS Report R42694, The Trans-Pacific Partnership (TPP) Negotiations and Issues for Congress, coordinated by Ian F. Fergusson.

[3] Many economists contend that using nominal exchange rates to convert foreign currency into U.S. dollars for comparing gross domestic product (GDP) may not be the most accurate measurement because prices vary from country to country. Purchasing power parity (PPP)

factors in price differences to reflect the actual purchasing power of currencies relative to the dollar in real terms.

[4] The World Bank utilizes a method for classifying world economies based on gross national product (GNP). Mexico is one of 48 economies classified as upper-middle-income, or countries which have a per capita GNP of \$3,946 to \$12,195 per year. The United States is one of 69 economies classified as a high-income, or countries which have a per capita GNP of more than \$12,195 per year.

[5] U.S. Bureau of Economic Analysis interactive statistics, available at http://www.bea.gov.

[6] Based on data from Global Trade Atlas.

[7] Mexico's export-oriented industries began with the maquiladora program established in the 1960s by the Mexican government, which allowed foreign-owned businesses to set up assembly plants in Mexico to produce for export. Maquiladoras could import intermediate materials duty-free with the condition that 20% of the final product be exported. The percentage of sales allowed to the domestic market increased over time as Mexico liberalized its trade regime. U.S. tariff treatment of maquiladora imports played a significant role in the industry. Under HTS provisions 9802.00.60 and 9802.00.80, the portion of an imported good that was of U.S. origin entered the United States duty-free. Duties were assessed only on the value added abroad. After NAFTA, North American rules of origin determine duty-free status. Recent changes in Mexican regulations on export-oriented industries merged the maquiladora industry and Mexican domestic assembly-for-export plants into one program called the Maquiladora Manufacturing Industry and Export Services (IMMEX).

[8] Inter-American Development Bank (IDB), "Mexico and Remittances," March 16, 2010, available at http://www.iadb.org/mif.

[9] See http://www.banxico.org.mx.

[10] IDB, Multilateral Investment Fund, Remittances to Latin America and the Caribbean in 2012: Differing Behavior Across Subregions, 2012.

[11] Federal Reserve Bank of Dallas, "Workers' Remittances to Mexico," El Paso Business Frontier, 2004.

[12] Inter-American Development Bank, "Mexico and Remittances," 2010.

[13] The White House, Office of the Press Secretary, "Fact Sheet: U.S.-Mexico High Level Economic Dialogue," September 20, 2013.

[14] Ibid.

[15] Department of Commerce, International Trade Administration, U.S.-Mexico High Level Regulatory Cooperation Council, http://www.trade.gov/hlrcc/.

[16] U.S. Department of State, Bureau of Western Hemisphere Affairs, United States-Mexico Partnership: Managing our 21st Century Border, Fact Sheet, April 29, 2013.

[17] For a fuller discussion of the 21st Century Border initiative, see: CRS Report R41349, U.S.-Mexican Security Cooperation: The Mérida Initiative and Beyond, by Clare Ribando Seelke and Kristin Finklea.

[18] U.S. Department of Homeland Security, 21st Century Border: The Executive Steering Committee, http://www.dhs.gov/executive-steering-committee.

[19] The White House, Office of the Press Secretary, Fact Sheet: Key Deliverables for the 2014 North American Leaders Summit, February 19, 2014.

[20] The White House, Office of the Press Secretary, "Joint Statement by North American Leaders—21st Century North America: Building the Most Competitive and Dynamic Region in the World," February 19, 2014.

[21] Rafael Amiel, Country Reports: Country Outlook: Economic - Mexico, IHS Connect, April 20, 2015.
[22] Ibid.
[23] Ibid.
[24] EIU, Mexico: Estimating the Fiscal Shortfall from Weak Oil, January 30, 2015.
[25] Ibid.
[26] Organization for Economic Cooperation and Development (OECD), OECD Economic Surveys, Mexico, January 2015, pp. 4-5.
[27] The World Bank, A Model from Mexico for the World, World Bank News Feature Story, November 19, 2014.
[28] For more information, see the Mexican government website: Secretaría de Desarrollo Social, Prospera Programa de Inclusión Social, at http://www.prospera.gob.mx.
[29] Gordon H. Hanson, p. 6.
[30] Ibid., p. 7.
[31] OECD 2015, p. 8.
[32] Ibid, p. 9.
[33] See CRS Report R42917, Mexico: Background and U.S. Relations, by Clare Ribando Seelke.
[34] Eduardo Bolio, Jaana Remes, and Tomas Lajous, et al., A Tale of Two Mexicos: Growth and Prosperity in a Two-Speed Economy, McKinsey Global Institute, March 2014.
[35] Ibid.
[36] Ibid., p. 2.
[37] See CRS Report R43313, Mexico's Oil and Gas Sector: Background, Reform Efforts, and Implications for the United States, coordinated by Clare Ribando Seelke.
[38] U.S. Energy Information Administration (EIA), Country Analysis Briefs: Mexico, April 24, 2014.
[39] Ibid.
[40] Adam Williams, "Mexico's Oil May Offer Gusher for Foreigners," The Washington Post, June 8, 2014, p. G3.
[41] EIA, April 24, 2014.
[42] "Richard Fausset, "Tons of Thousands Protest Mexican Oil Reforms," Los Angeles Times, September 8, 2013.
[43] Enrique Krauze, "Mexico's Theology of Oil," New York Times, November 1, 2013.
[44] Party of the Democratic Revolution (Partido de la Revolución Democrática; PRD), "National Position of the PRD and its Parliamentary Groups in the Senate and the Chamber of Deputies regarding the Secondary Legislation on Energy Matters," June 10, 2014.
[45] U.S. Department of State, "Remarks by Secretary of State Hillary Rodham Clinton at the Signing of the U.S.-Mexico Transboundary Agreement," February 20, 2012.
[46] The WTO allows member countries to form regional trade agreements under Article under certain rules. The position of the WTO is that regional trade agreements can often support the WTO's multilateral trading system by allowing groups of countries to negotiate rules and commitments that go beyond what was possible at the time under the WTO. The WTO has a committee on regional trade agreements that examines regional groups and assesses whether they are consistent with WTO rules. See The World Trade Organization, "Understanding the WTO: Cross-Cutting and New Issues, Regionalism: Friends or Rivals?" http://www.wto.org.
[47] See CRS Report R43748, The Pacific Alliance: A Trade Integration Initiative in Latin America, by M. Angeles Villarreal

[48] See CRS Report R42965, The North American Free Trade Agreement (NAFTA), by M. Angeles Villarreal and Ian F. Fergusson.

[49] See CRS Report R41660, U.S.-South Korea Free Trade Agreement and Potential Employment Effects: Analysis of Studies, by Mary Jane Bolle and James K. Jackson.

[50] See, for example, Gary Clyde Hufbauer and Jeffrey J. Schott, NAFTA Revisited: Achievements and Challenges, Institute for International Economics, October 2005; Center for Strategic and International Studies, NAFTA's Impact on North America: The First Decade, Edited by Sidney Weintraub, 2004; and U.S. Chamber of Commerce, Opening Markets, Creating Jobs: Estimated U.S. Employment Effects of Trade with FTA Partners, 2010.

[51] See, for example, Robert E. Scott, Heading South: U.S.-Mexico Trade and Job Displacement under NAFTA, Economic Policy Institute, May 3, 2011; and the Frederick S. Pardee Center, The Future of North American Trade Policy: Lessons from NAFTA, Boston University, November 2009.

[52] Hufbauer and Schott, NAFTA Revisited, pp. 20-21.

[53] Ibid., p. 21.

[54] Robert Koopman, William Powers, and Zhi Wang, et al., Give Credit Where Credit is Due: Tracing Value Added in Global Production Chains, National Bureau of Economic Research, Working Paper 16426, Cambridge, MA, September 2010, pp. 7-8.

[55] See CRS Report R42694, The Trans-Pacific Partnership (TPP) Negotiations and Issues for Congress, coordinated by Ian F. Fergusson.

[56] See CRS Report R43491, Trade Promotion Authority (TPA): Frequently Asked Questions, by Ian F. Fergusson and Richard S. Beth, and CRS Report IF10038, Trade Promotion Authority (TPA), by Ian F. Fergusson.

[57] Emily Pickrell and Jerome Ashton, "Mexico Focusing on Trans-Pacific Partnership for Market Access," Bloomberg BNA, January 7, 2015.

[58] See CRS In Focus IF10034, New Era Dawns in U.S.-Mexico Sugar Trade, by Mark A. McMinimy.

[59] Rosella Brevetti, "Two Companies Step Up Attack on Deals Commerce Negotiated on Sugar From Mexico," Bloomberg BNA, January 20, 2015.

[60] Emily Pickrell, "Mexican Trade Official Praises ITC Decision Upholding Suspension of Sugar Investigations," Bloomberg BNA, March 24, 2015.

[61] See International Trade Administration, Fact Sheet: Commerce Initiates Antidumping Duty and Countervailing Duty Investigations of Imports of Sugar from Mexico, at http://enforcement.trade.gov/download/factsheets/factsheet-mexicosugar-cvd-initiation-041814.pdf.

[62] U.S. International Trade Commission, Sugar from Mexico, Investigation Nos. 701-TA-513 and 731-TA-1249 (Preliminary), Publication 4467, Washington, DC, May 2014, p. 3.

[63] CRS In Focus IF10034, New Era Dawns on U.S.-Mexico Sugar Trade, by Mark A. McMinimy.

[64] Sweetener Users Association, "SUA Statement on Commerce Department's Postponement of Preliminary Antidumping Duty Determination," press release, August 21, 2014, http://sweetenerusers.org.

[65] Ibid.

[66] "Commerce Finds Countervailable Subsidies in Mexican Sugar Trade Case," World Trade Online, August 25, 2014.

[67] See "Bush Administration Defends Sugar Deal to Congress," Inside U.S. Trade, November 3, 2006; "Grassley, U.S. Industry Welcome Agreement with Mexico on Sugar, HFCS,"

International Trade Reporter, August 3, 2006; and, "U.S., Mexico Reach Agreement on WTO Soft Drink Dispute Compliance Deadline," International Trade Reporter, July 13, 2006.

[68] Mexico Ministry of the Economy, Impact of Country of Origin Labeling (COOL) in the U.S.-Mexico Trade Partnership, March 2015.

[69] Heather Rothman, "Agriculture Lawmakers, Meat Industry Urge Repeal of COOL to Solve Trade Dispute," Bloomberg BNA, March 25, 2015.

[70] World Trade Organization, United States-Certain Country of Origin Labelling Requirements, Dispute Settlement: Dispute DS384, http://www.wto.org.

[71] For more information, see CRS Report RS22955, Country-of-Origin Labeling for Foods and the WTO Trade Dispute on Meat Labeling, by Joel L. Greene.

[72] Rosella Brevetti, "Labeling Dispute Casts Shadow of Possible Retaliation on U.S. Exports in 2015," Bloomberg BNA, January 7, 2015.

[73] Federal Motor Carrier Safety Administration (FMCSA), United States to Expand Trade Opportunities with Mexico through Safe Cross-Border Trucking, January 9, 2015, available at http://www.fmcsa.dot.gov.

[74] Emily Pickrell, "Mexico Plans to Retaliate if Lawsuit Closes Doors to Cross-Border Trucking," Bloomberg BNA, March 11, 2015.

[75] See CRS Report R41821, Status of Mexican Trucks in the United States: Frequently Asked Questions, by John Frittelli.

[76] Rosella Brevetti, "Transportation Department Clears Way of Roadlblocks for Mexican Motor Carriers," January 9, 2015.

[77] FMCSA, The Cross-Border Trucking Pilot Program Report to Congress, available at: http://www.fmcsa.dot.gov/mission/policy/cross-border-trucking-pilot-program-report-congress.

[78] Rosella Brevetti, "Transportation Department Clears Way of Roadlblocks for Mexican Motor Carriers," January 9, 2015.

[79] Ibid.

[80] Department of Transportation, "Cross-Border Trucking Demonstration Project," March 11, 2008.

[81] Rosella Brevetti, "Key GOP House Members Urge Obama to Develop New Mexico Truck Program," International Trade Reporter, March 26, 2009.

[82] Amy Tsui, "Plan to Resolve Mexican Trucking Dispute 'Very Near,' DOT's LaHood Tells Lawmakers," International Trade Reporter, March 11, 2010.

[83] Ibid.

[84] Inside U.S. Trade's World Trade Online, "Pork, Cheeses, Fruits to Face new Tariffs Due to Mexico Trucks Dispute," August 17, 2010.

[85] Inside U.S. Trade's World Trade Online, "New Mexican Retaliatory Tariffs in Trucks Dispute Designed to Spur U.S.," September 3, 2010.

[86] Washington Trade Daily, "A Trucking Breakthrough," Volume 20, No. 45, March 4, 2011.

[87] Rosella Brevetti and Nacha Cattan, "DOT's LaHood Presents 'Concept' Paper on Resolving NAFTA Mexico Truck Dispute," January 13, 2011.

[88] U.S. Department of Transportation, "U.S. Cross-Border Trucking Effort Emphasizes Safety and Efficiency," Press Release, January 6, 2011.

[89] Josh Mitchell, "U.S. Jump-Starts Bid to End Truck Dispute with Mexico," Wall Street Journal, January 7, 2011.

[90] U.S. Department of Transportation, Concept Document: Phased U.S.-Mexico Cross-Border Long Haul Trucking Proposal, January 6, 2011, at http://www.fmcsa.dot.gov.

[91] Federal Motor Carrier Safety Administration (FMCSA), "United States and Mexico Announce Safe, Secure Cross-Border Trucking Program: U.S.-Mexico Agreements Will Lift Tariffs and Put Safety First," News Release, July 6, 2011.

[92] NAFTA Works, "The United States and Mexico Sign a Memorandum of Understanding on Long-Hayl Cross-Border Trucking," Volume 3, Alert 18, July 2011.

[93] FMCSA, "United States and Mexico Announce Safe, Secure Cross-Border Trucking Program: U.S.-Mexico Agreements Will Lift Tariffs and Put Safety First," July 6, 2011.

[94] Ibid.

[95] Ibid.

[96] Rosella Brevetti, "Mexico Suspends Tariffs as Trucking Program is Launched," International Trade Reporter, October 27, 2011.

[97] Federal Motor Carrier Safety Administration (FMCSA), United States to Expand Trade Opportunities with Mexico through Safe Cross-Border Trucking, January 9, 2015, available at http://www.fmcsa.dot.gov.

[98] Stephanie Strom, "United States and Mexico Reach Tomato Deal, Averting a Trade War," New York Times, February 4, 2013.

[99] U.S. Department of Commerce, Import Administration, Fresh Tomatoes from Mexico 1996 Suspension Agreement, available at http://ia.ita.doc.gov/tomato/index.html.

[100] Rosella Brevetti, "Mexico Ambassador Warns Against ending U.S. Suspension Agreement on Tomatoes," International Trade Reporter, September 20, 2012.

[101] Brian Flood, "Florida Tomato Trade Group Loses Bid to Block Evidence in Mexico Dispute," Bloomberg BNA, April 15, 2014.

[102] U.S. Department of Commerce, Import Administration, Fresh Tomatoes from Mexico 1996 Suspension Agreement, available at http://ia.ita.doc.gov/tomato/index.html.

[103] Ibid.

[104] Len Bracken, "Commerce, Mexican Tomato Growers Agree on Final Version of Antidumping Agreement," International Trade Daily, March 5, 2013.

[105] Inside U.S. Trade's World Trade Online, "U.S. Growers Seek to End Suspension Agreement on Mexican Tomato Imports," June 28, 2012.

[106] Stephanie Strom, New York Times, February 4, 2013.

[107] Office of the United States Trade Representative (USTR), "U.S. Appeal in WTO Dolphin-Safe Tuna Labeling Dispute with Mexico," January 23, 2012.

[108] Bryce Baschuk, "Mexico Prevails in Latest WTO Dispute over U.S. Tuna Labeling Rules, Bloomberg BNA, April 14, 2015.

[109] Ibid.

[110] Ibid. For more information, see the USTR website at http://www.ustr.gov.

[111] Daniel Pruzin, "Appellate Body Overturns Key Panel Findings on U.S. Tuna-Dolphin Labeling Requirements," International Trade Reporter, May 24, 2012.

[112] For more information, see World Trade Organization, United States—Measures Concerning the Importation, Marketing, and Sale of Tuna and Tuna Products, available at http://www.wto.org.

[113] Tim Carman, "Tuna, meat labeling disputes highlight WTO control," Washington Post, January 10, 2012.

[114] Ibid.

[115] See CRS Report R42344, Trans-Pacific Partnership (TPP) Countries: Comparative Trade and Economic Analysis, by Brock R. Williams.

INDEX

D

G

H

I

J

K

L

M

N

O

P

Q

R

U

V

W